INTERNATIONAL
MATHEMATICAL OLYMPIADS, 1978–1985
AND
FORTY SUPPLEMENTARY PROBLEMS

NEW MATHEMATICAL LIBRARY

PUBLISHED BY

THE MATHEMATICAL ASSOCIATION OF AMERICA

The New Mathematical Library (NML) was started in 1961 by the School Mathematics Study Group to make available to high school students short expository books on various topics not usually covered in the high school syllabus. In a decade the NML matured into a steadily growing series of some twenty titles of interest not only to the originally intended audience, but to college students and teachers at all levels. Previously published by Random House and L. W. Singer, the NML became a publication series of the Mathematical Association of America (MAA) in 1975. Under the auspices of the MAA the NML continues to grow and remains dedicated to its original and expanded purposes. In its third decade, it contains some thirty titles.

INTERNATIONAL
MATHEMATICAL OLYMPIADS, 1978–1985
AND
FORTY SUPPLEMENTARY
PROBLEMS

Compiled and with solutions by
Murray S. Klamkin

University of Alberta

31

THE MATHEMATICAL ASSOCIATION
OF AMERICA

Second Printing
© 1986 by the Mathematical Association of America (Inc.)
Published in Washington, D.C. by
The Mathematical Association of America

Publication of this book was
generously supported by UNESCO
through its Division of Science, Technical,
and Environmental Education

Library of Congress Catalog Card Number 86-061792

Complete Set ISBN: 0-88385-600-X

Vol. 31 ISBN: 0-88386-631-X

Manufactured in the United States of America

To my wife

Irene

Without her patience and assistance this book would still be in limbo.

NEW MATHEMATICAL LIBRARY

Other titles in preparation.

Editors' Note

The Mathematical Association of America is pleased to publish this sequel to NML vol. 27; it contains the International Mathematical Olympiads, 1978–1985, and forty supplementary problems selected by Murray S. Klamkin, who prepared this text. The educational impact of such problems in stimulating mathematical thinking of young students and its long range effects have been eloquently described, both from the viewpoint of the participant and that of the mature mathematician in retrospect, by Gábor Szegö in his preface to the Hungarian Problem Books, volumes 11 and 12 of this NML series.

Our aim in all problem collections of this series is not only to help the high school student satisfy his curiosity by presenting solutions with tools familiar to him, but also to instruct him in the use of more sophisticated methods and different modes of attack by including explanatory material and alternate solutions. For problem solvers, each problem is a challenging entity to be conquered; for theory spinners, each problem is the proof of their pudding. It is the fruitful synthesis of these seemingly antithetical forces that we have tried to achieve.

We are extremely grateful to Murray S. Klamkin, the dedicated coach and expert problem solver, for having compiled the solutions to all the problems. The editors enjoyed their task and gratefully acknowledge Peter Ungar's elaborations, addenda and applications of some solutions.

The editors of the present collection have occasionally departed somewhat from the wording of the problems originally presented to the English-speaking contestants. This was done in the interest of clarity and smooth style; since translations from one language into another are seldom completely faithful, we felt that such small departures were permissible.

We close this foreword by quoting G. Szegö's concluding observation from his preface to NML volumes 11 and 12:

"We should not forget that the solution of any worthwhile problem very rarely comes to us easily and without hard work; it is rather the result of intellectual effort of days or weeks or months. Why should the young mind be willing to make this supreme effort? The explanation is probably the instinctive preference for certain values, that is, the attitude which rates intellectual effort and spiritual achievement higher than material advantage.

Such a valuation can be only the result of a long cultural development of environment and public spirit which is difficult to accelerate by governmental aid or even by more intensive training in mathematics. The most effective means may consist of transmitting to the young mind the beauty of intellectual work and the feeling of satisfaction following a great and successful mental effort. The hope is justified that the present book might aid exactly in this respect and that it represents a good step in the right direction."

Donald J. Albers
William G. Chinn
Joanne Elliott
Basil Gordon
Ivan Niven
Max Schiffer
Anneli Lax

September 1986

Preface

This volume is a sequel to "International Mathematical Olympiads 1959–1977", NML vol. 27, by S. L. Greitzer. For a history of the IMO and various national olympiads, I refer the reader to the preface of NML vol. 27, to "ICMI Report on the Mathematical Contests in Secondary Education I" edited by Hans Freudenthal in *Educational Studies in Mathematics 2* (1969) pp. 80–114, and to my Olympiad Corner #1–#70, in Crux Mathematicorum 1979–1985, in particular #3, #4, #8, #58 and #68. In the present volume, I have included information on the rules governing the IMO (Appendix A) and a tabulation of participating teams and their scores (Appendix B).

This sequel consists of all the problems and solutions in the IMO's from the Twentieth IMO (1978) through the Twenty-Sixth IMO (1985). To round out the collection, I have included a set of problems proposed for the IMO's over the last several years, but not selected, and their solutions. As noted in Appendix A, at the end of every IMO, the delegates of each participating country are provided with copies of the submitted proposed problems together with their solutions, if available. These are usually given in the four official languages: English, French, German and Russian. Each year, quite a number of the participating countries publish the problems and solutions of the IMO together with those of their national olympiad in either a special olympiad report, or a journal or a book. In the USA, these appear in a pamphlet available from the MAA as well as in Mathematics Magazine. Naturally, most of the solutions are similar to the original, submitted solutions or to the official solutions, and the same is true of this collection. Here, however, solutions have been edited, quite a number have been changed and/or expanded, and references have been added where pertinent. There is a comprehensive bibliography at the end, with items numbered and referred to in brackets in the text.

A frequent concern of contestants is how detailed a solution has to be to obtain full marks. This, of course, depends on the graders. However, when in doubt, one should include details rather than leave them out. The solutions given here are more detailed than need be for the IMO contestants; it was the consensus of the NML Committee that this would make them more accessible and of service to a wider audience.

In the IMO there are special prizes for elegant solutions and/or non-trivial generalizations with proof. Although the elegant solutions are much more satisfactory and transparent than the non-elegant ones, finding them usually takes a longer time, unless one has some special à priori knowledge, or else one has had long practise in finding them. So contestants should not "go too far out of their way" looking for elegance in an IMO. Any non-elegant correct solution will still receive full marks. Nevertheless, if there is time, contestants are advised to strive for refinements, since elegance is frequently a sign of real understanding.

No doubt, many of the solutions given here can be improved upon or generalized; when a reader finds this to be the case, I would be very grateful to receive any such communications.

Many thanks go to Samuel L. Greitzer for sharing the joys and burdens of coaching the USA Olympiad teams from 1975–1980, and to Andy Liu from 1981–1984. I am also grateful to Samuel L. Greitzer, Peter D. Lax, Andy Liu, Amram Meir, Greg Patruno, Jim Pounder, Cecil Rousseau, Peter Ungar and all the members of the NML Committee, particularly Joanne Elliott, Basil Gordon, Anneli Lax and Ivan Niven, for many improvements and additions.

Murray S. Klamkin
University of Alberta

Contents

Problems

Twentieth International Olympiad, 1978†

1978/1. m and n are natural numbers with $1 \leqslant m < n$. In their decimal representations, the last three digits of 1978^m are equal, respectively, to the last three digits of 1978^n. Find m and n such that $m + n$ has its least value.

1978/2. P is a given point inside a given sphere. Three mutually perpendicular rays from P intersect the sphere at points U, V, and W; Q denotes the vertex diagonally opposite to P in the parallelepiped determined by PU, PV, and PW. Find the locus of Q for all such triads of rays from P.

1978/3. The set of all positive integers is the union of two disjoint subsets $\{f(1), f(2), \ldots, f(n), \ldots\}$, $\{g(1), g(2), \ldots, g(n), \ldots\}$, where

$$f(1) < f(2) < \cdots < f(n) < \cdots,$$
$$g(1) < g(2) < \cdots < g(n) < \cdots,$$

and

$$g(n) = f(f(n)) + 1 \qquad \text{for all} \quad n \geqslant 1.$$

Determine $f(240)$.

1978/4. In triangle ABC, $AB = AC$. A circle is tangent internally to the circumcircle of triangle ABC and also to sides AB, AC at P, Q, respectively. Prove that the midpoint of segment PQ is the center of the incircle of triangle ABC.

---

†Solutions of the 1978 problems begin on p. 15.

1978/5. Let $\{a_k\}$ $(k = 1, 2, 3, \ldots, n, \ldots)$ be a sequence of distinct positive integers. Prove that for all natural numbers n,

$$\sum_{k=1}^{n} \frac{a_k}{k^2} \geqslant \sum_{k=1}^{n} \frac{1}{k}.$$

1978/6. An international society has its members from six different countries. The list of members contains 1978 names, numbered $1, 2, \ldots, 1978$. Prove that there is at least one member whose number is the sum of the numbers of two members from his own country, or twice as large as the number of one member from his own country.

Twenty-first International Olympiad, 1979†

1979/1. Let p and q be natural numbers such that

$$\frac{p}{q} = 1 - \frac{1}{2} + \frac{1}{3} - \frac{1}{4} + \cdots - \frac{1}{1318} + \frac{1}{1319}.$$

Prove that p is divisible by 1979.

1979/2. A prism with pentagons $A_1A_2A_3A_4A_5$ and $B_1B_2B_3B_4B_5$ as top and bottom faces is given. Each side of the two pentagons and each of the line-segments A_iB_j, for all $i, j = 1, \ldots, 5$, is colored either red or green. Every triangle whose vertices are vertices of the prism and whose sides have all been colored has two sides of a different color. Show that all 10 sides of the top and bottom faces are the same color.

1979/3. Two circles in a plane intersect. Let A be one of the points of intersection. Starting simultaneously from A two points move with constant speeds, each point travelling along its own circle in the same sense. The two points return to A simultaneously after one revolution. Prove that there is a fixed point P in the plane such that, at any time, the distances from P to the moving points are equal.

1979/4. Given a plane π, a point P in this plane and a point Q not in π, find all points R in π such that the ratio $(QP + PR)/QR$ is a maximum.

---

†Solutions of the 1979 problems begin on p. 23.

1979/5. Find all real numbers a for which there exist non-negative real numbers x_1, x_2, x_3, x_4, x_5 satisfying the relations

$$\sum_{k=1}^{5} kx_k = a, \qquad \sum_{k=1}^{5} k^3 x_k = a^2, \qquad \sum_{k=1}^{5} k^5 x_k = a^3.$$

1979/6. Let A and E be opposite vertices of a regular octagon. A frog starts jumping at vertex A. From any vertex of the octagon except E, it may jump to either of the two adjacent vertices. When it reaches vertex E, the frog stops and stays there. Let a_n be the number of distinct paths of exactly n jumps ending at E. Prove that $a_{2n-1} = 0$,

$$a_{2n} = \frac{1}{\sqrt{2}}(x^{n-1} - y^{n-1}), \qquad n = 1, 2, 3, \ldots,$$

where $x = 2 + \sqrt{2}$ and $y = 2 - \sqrt{2}$.

Note. A path of n jumps is a sequence of vertices $(P_0, \ldots, P_n)$ such that

(i) $P_0 = A$, $P_n = E$;

(ii)   for every i, $0 \leqslant i \leqslant n - 1$, P_i is distinct from E;

(iii)   for every i, $0 \leqslant i \leqslant n - 1$, P_i and P_{i+1} are adjacent.

Twenty-second International Olympiad, 1981†

1981/1. P is a point inside a given triangle ABC. D, E, F are the feet of the perpendiculars from P to the lines BC, CA, AB respectively. Find all P for which

$$\frac{BC}{PD} + \frac{CA}{PE} + \frac{AB}{PF}$$

is least.

1981/2. Let $1 \leqslant r \leqslant n$ and consider all subsets of r elements of the set $\{1, 2, \ldots, n\}$. Each of these subsets has a smallest member. Let $F(n, r)$ denote the arithmetic mean of these smallest numbers; prove that

$$F(n, r) = \frac{n+1}{r+1}.$$

†Solutions of the 1981 problems begin on p. 30.

1981/3. Determine the maximum value of $m^2 + n^2$, where m and n are integers satisfying $m, n \in \{1, 2, \ldots, 1981\}$ and $(n^2 - mn - m^2)^2 = 1$.

1981/4. (a) For which values of $n > 2$ is there a set of n consecutive positive integers such that the largest number in the set is a divisor of the least common multiple of the remaining $n - 1$ numbers?
  (b) For which values of $n > 2$ is there exactly one set having the stated property?

1981/5. Three congruent circles have a common point O and lie inside a given triangle. Each circle touches a pair of sides of the triangle. Prove that the incenter and the circumcenter of the triangle and the point O are collinear.

1981/6. The function $f(x, y)$ satisfies

(1) $$f(0, y) = y + 1,$$

(2) $$f(x + 1, 0) = f(x, 1),$$

(3) $$f(x + 1, y + 1) = f(x, f(x + 1, y)),$$

for all non-negative integers x, y. Determine $f(4, 1981)$.

Twenty-third International Olympiad, 1982†

1982/1. The function $f(n)$ is defined for all positive integers n and takes on non-negative integer values. Also, for all m, n

$$f(m + n) - f(m) - f(n) = 0 \text{ or } 1$$

$$f(2) = 0, \quad f(3) > 0, \quad \text{and} \quad f(9999) = 3333.$$

Determine $f(1982)$.

1982/2. A non-isosceles triangle $A_1 A_2 A_3$ is given with sides a_1, a_2, a_3 (a_i is the side opposite A_i). For all $i = 1, 2, 3$, M_i is the midpoint of side a_i, and T_i is the point where the incircle touches side a_i. Denote by S_i the reflection of T_i in the interior bisector of angle A_i. Prove that the lines $M_1 S_1$, $M_2 S_2$, and $M_3 S_3$ are concurrent.

†Solutions of the 1982 problems begin on p. 37.

1982/3. Consider the infinite sequences $\{x_n\}$ of positive real numbers with the following properties:

$$x_0 = 1, \quad \text{and for all } i \geqslant 0, \quad x_{i+1} \leqslant x_i.$$

(a) Prove that for every such sequence, there is an $n \geqslant 1$ such that

$$\frac{x_0^2}{x_1} + \frac{x_1^2}{x_2} + \cdots + \frac{x_{n-1}^2}{x_n} \geqslant 3.999.$$

(b) Find such a sequence for which

$$\frac{x_0^2}{x_1} + \frac{x_1^2}{x_2} + \cdots + \frac{x_{n-1}^2}{x_n} < 4 \qquad \qquad \text{for all} \quad n.$$

1982/4. Prove that if n is a positive integer such that the equation

$$x^3 - 3xy^2 + y^3 = n$$

has a solution in integers (x, y), then it has at least three such solutions. Show that the equation has no solutions in integers when $n = 2891$.

1982/5. The diagonals AC and CE of the regular hexagon $ABCDEF$ are divided by the inner points M and N, respectively, so that

$$\frac{AM}{AC} = \frac{CN}{CE} = r.$$

Determine r if B, M, and N are collinear.

1982/6. Let S be a square with sides of length 100, and let L be a path within S which does not meet itself and which is composed of line segments $A_0A_1, A_1A_2, \ldots, A_{n-1}A_n$ with $A_0 \neq A_n$. Suppose that for every point P of the boundary of S there is a point of L at a distance from P not greater than $1/2$. Prove that there are two points X and Y in L such that the distance between X and Y is not greater than 1, and the length of that part of L which lies between X and Y is not smaller than 198.

Twenty-fourth International Olympiad, 1983†

1983/1. Find all functions f defined on the set of positive real numbers which take positive real values and satisfy the conditions:

(i) $f(xf(y)) = yf(x)$ for all positive x, y;

(ii) $f(x) \to 0$ as $x \to \infty$.

†Solutions of the 1983 problems begin on p. 43.

1983/2. Let A be one of the two distinct points of intersection of two unequal coplanar circles C_1 and C_2 with centers O_1 and O_2, respectively. One of the common tangents to the circles touches C_1 at P_1 and C_2 at P_2, while the other touches C_1 at Q_1 and C_2 at Q_2. Let M_1 be the midpoint of P_1Q_1 and M_2 be the midpoint of P_2Q_2. Prove that $\angle O_1AO_2 = \angle M_1AM_2$.

1983/3. Let a, b and c be positive integers, no two of which have a common divisor greater than 1. Show that $2abc - ab - bc - ca$ is the largest integer which cannot be expressed in the form $xbc + yca + zab$, where x, y and z are non-negative integers.

1983/4. Let ABC be an equilateral triangle and $\mathcal{E}$ the set of all points contained in the three segments AB, BC and CA (including A, B and C). Determine whether, for every partition of $\mathcal{E}$ into two disjoint subsets, at least one of the two subsets contains the vertices of a right-angled triangle. Justify your answer.

1983/5. Is it possible to choose 1983 distinct positive integers, all less than or equal to 10^5, no three of which are consecutive terms of an arithmetic progression? Justify your answer.

1983/6. Let a, b and c be the lengths of the sides of a triangle. Prove that

$$a^2b(a - b) + b^2c(b - c) + c^2a(c - a) \geqslant 0.$$

Determine when equality occurs.

Twenty-fifth International Olympiad, 1984†

1984/1. Prove that $0 \leqslant yz + zx + xy - 2xyz \leqslant 7/27$, where x, y and z are non-negative real numbers for which $x + y + z = 1$.

1984/2. Find one pair of positive integers a and b such that:

(i) $ab(a + b)$ is not divisible by 7;

(ii) $(a + b)^7 - a^7 - b^7$ is divisible by 7^7.

Justify your answer.

†Solutions of the 1984 problems begin on p. 48.

1984/3. In the plane two different points O and A are given. For each point X of the plane, other than O, denote by $a(X)$ the measure of the angle between OA and OX in radians, counterclockwise from OA $(0 \leqslant a(X) < 2\pi)$. Let $C(X)$ be the circle with center O and radius of length $OX + a(X)/OX$. Each point of the plane is colored by one of a finite number of colors. Prove that there exists a point Y for which $a(Y) > 0$ such that its color appears on the circumference of the circle $C(Y)$.

1984/4. Let $ABCD$ be a convex quadrilateral such that the line CD is a tangent to the circle on AB as diameter. Prove that the line AB is a tangent to the circle on CD as diameter if and only if the lines BC and AD are parallel.

1984/5. Let d be the sum of the lengths of all the diagonals of a plane convex polygon with n vertices $(n > 3)$, and let p be its perimeter. Prove that

$$n - 3 < \frac{2d}{p} < \left[\frac{n}{2}\right]\left[\frac{n+1}{2}\right] - 2,$$

where $[x]$ denotes the greatest integer not exceeding x.

1984/6. Let a, b, c and d be odd integers such that $0 < a < b < c < d$ and $ad = bc$. Prove that if $a + d = 2^k$ and $b + c = 2^m$ for some integers k and m, then $a = 1$.

Twenty-sixth International Olympiad, 1985†

1985/1. A circle has center on the side AB of the cyclic quadrilateral $ABCD$. The other three sides are tangent to the circle. Prove that $AD + BC = AB$.

1985/2. Let n and k be given relatively prime natural numbers, $k < n$. Each number in the set $M = \{1, 2, \ldots, n - 1\}$ is colored either blue or white. It is given that
  (i)   for each $i \in M$, both i and $n - i$ have the same color;
  (ii) for each $i \in M$, $i \neq k$, both i and $|i - k|$ have the same color.
Prove that all numbers in M must have the same color.

---

†Solutions of the 1985 problems begin on p. 54.

1985/3. For any polynomial $P(x) = a_0 + a_1x + \cdots + a_kx^k$ with integer coefficients, the number of coefficients which are odd is denoted by $w(P)$. For $i = 0, 1, \ldots,$ let $Q_i(x) = (1 + x)^i$. Prove that if $i_1, i_2, \ldots, i_n$ are integers such that $0 \leqslant i_1 < i_2 < \cdots < i_n$, then

$$w(Q_{i_1} + Q_{i_2} + \cdots + Q_{i_n}) \geqslant w(Q_{i_1}).$$

1985/4. Given a set M of 1985 distinct positive integers, none of which has a prime divisor greater than 26. Prove that M contains at least one subset of four distinct elements whose product is the fourth power of an integer.

1985/5. A circle with center O passes through the vertices A and C of triangle ABC and intersects the segments AB and BC again at distinct points K and N, respectively. The circumscribed circles of the triangles ABC and KBN intersect at exactly two distinct points B and M. Prove that angle OMB is a right angle.

1985/6. For every real number x_1, construct the sequence $x_1, x_2, \ldots$ by setting

$$x_{n+1} = x_n\left(x_n + \frac{1}{n}\right) \qquad \text{for each} \quad n \geqslant 1.$$

Prove that there exists exactly one value of x_1 for which

$$0 < x_n < x_{n+1} < 1$$

for every n.

Supplementary Problems

Algebra†

A/1. Let n be a positive integer having at least two distinct prime factors. Show that there is a permutation $(a_1, a_2, \ldots, a_n)$ of $(1, 2, \ldots, n)$ such that

$$\sum_{k=1}^{n} k \cos \frac{2\pi a_k}{n} = 0.$$

A/2. A polynomial $P(x)$ of degree 990 satisfies

$$P(k) = F_k, \qquad k = 992, 993, \ldots, 1982,$$

where $\{F_k\}$ is the Fibonacci sequence, defined by

$$F_1 = F_2 = 1, \qquad F_{n+1} = F_n + F_{n-1}, \qquad n = 2, 3, 4 \ldots .$$

Prove that $P(1983) = F_{1983} - 1$.

A/3. Prove that there is a unique infinite sequence $\{u_0, u_1, u_2, \ldots\}$ of positive integers such that, for all $n \geqslant 0$,

$$u_n^2 = \sum_{r=0}^{n} \binom{n+r}{r} u_{n-r}.$$

A/4. Let a, b and c be real numbers such that

$$(bc - a^2)^{-1} + (ca - b^2)^{-1} + (ab - c^2)^{-1} = 0.$$

Prove that

$$a(bc - a^2)^{-2} + b(ca - b^2)^{-2} + c(ab - c^2)^{-2} = 0.$$

†Solutions of the supplementary algebra problems begin on p. 61.

A/5. Three roots of the equation

$$x^4 - px^3 + qx^2 - rx + s = 0$$

are $\tan A$, $\tan B$, $\tan C$, where A, B, C are the angles of a triangle. Determine the fourth root as a function of (only) p, q, r, and s.

A/6. Let $S_k = x_1^k + x_2^k + \cdots + x_n^k$, where the x_i are real numbers. If

$$S_1 = S_2 \cdots = S_{n+1},$$

prove that $x_i \in \{0, 1\}$ for every $i = 1, 2, \ldots, n$.

A/7. If the lengths a, b, c of the sides of a triangle satisfy

$$2(bc^2 + ca^2 + ab^2) = b^2c + c^2a + a^2b + 3abc,$$

prove that the triangle is equilateral. Prove also that the equation can be satisfied by positive real numbers that are not lengths of sides of a triangle.

A/8. Let $\{a_n\}$ and $\{b_n\}$, $n = 1, 2, 3, \ldots$, be two sequences of natural numbers such that, for all $n \geqslant 1$,

$$a_{n+1} = na_n + 1 \qquad \text{and} \qquad b_{n+1} = nb_n - 1.$$

Prove that the two sequences can have only a finite number of terms in common.

A/9. Determine all sequences $\{a_1, a_2, \ldots\}$ such that

$$a_1 = 1 \qquad \text{and} \qquad |a_n - a_m| \leqslant \frac{2mn}{m^2 + n^2}$$

for all positive integers m and n.

A/10. Determine all continuous functions f such that, for all real x and y,

$$f(x + y)f(x - y) = \{f(x)f(y)\}^2.$$

Number Theory†

N.T./1. Show that there exist infinitely many sets of 1983 consecutive positive integers each of which is divisible by some number of the form a^{1983}, where a is a positive integer $\neq 1$.

†Solutions of the supplementary number theory problems begin on p. 75.

N.T./2. Each of the numbers $x_1, x_2, \ldots, x_n$ equals 1 or -1, and

$$x_1 x_2 x_3 x_4 + x_2 x_3 x_4 x_5 + x_3 x_4 x_5 x_6 + \cdots + x_{n-3} x_{n-2} x_{n-1} x_n$$

$$+ x_{n-2} x_{n-1} x_n x_1 + x_{n-1} x_n x_1 x_2 + x_n x_1 x_2 x_3 = 0.$$

Prove that n is divisible by 4.

N.T./3. Factor the number $5^{1985} - 1$ into a product of three integers, each of which is larger than 5^{100}.

N.T./4. Let m boxes be given, with some balls in each box. Let $n < m$ be a given integer. The following operation is performed: choose n of the boxes and put 1 ball in each of them. Prove:
(a) If m and n are relatively prime, then it is possible, by performing the operation a finite number of times, to arrive at a situation where all the boxes contain an equal number of balls.
(b) If m and n are not relatively prime, there exist initial distributions of balls in the boxes such that an equal distribution is not possible to achieve.

N.T./5. If $(1 + x + x^2 + x^3 + x^4)^{496} = a_0 + a_1 x + \cdots + a_{1984} x^{1984}$,
(i) determine the g.c.d. of the coefficients $a_3, a_8, a_{13}, \ldots, a_{1983}$;
(ii) show that $10^{347} > a_{992} > 10^{340}$.

N.T./6. Let $a_0 = 0$ and

$$a_{n+1} = k(a_n + 1) + (k + 1)a_n + 2\sqrt{k(k + 1)a_n(a_n + 1)},$$

$$n = 0, 1, 2, \ldots,$$

where k is a positive integer. Prove that a_n is a positive integer for $n = 1, 2, 3, \ldots$.

Plane Geometry†

P.G./1. Let $A_1 A_2, B_1 B_2, C_1 C_2$ be three equal segments on the sides of an equilateral triangle. Prove that in the triangle formed by the lines $B_2 C_1, C_2 A_1, A_2 B_1$, the segments $B_2 C_1, C_2 A_1, A_2 B_1$ are proportional to the sides in which they are contained.

†Solutions of the supplementary plane geometry problems begin on p. 78.

P.G./2. In the plane, a circle with radius r and center O and a line l are given, and the distance from O to l is d, $d > r$. The points M and N are chosen on l in such a way that the circle with diameter MN is externally tangent to the given circle. Show that there exists a point A in the plane such that all the segments MN subtend a constant angle at A.

P.G./3. Determine whether or not there exist 100 distinct lines in the plane having exactly 1985 distinct points of intersection.

P.G./4. A closed convex set F lies inside a circle with center O. The angle subtended by F from every point of the circle is 90°. Prove that O is a center of symmetry of F.

Solid Geometry†

S.G./1. The altitude from a vertex of a given tetrahedron intersects the opposite face in its orthocenter. Prove that all four altitudes of the tetrahedron are concurrent.

S.G./2. The sum of all the face angles at all but one of the vertices of a given simple polyhedron is 5160°. Find the sum of all the face angles of the polyhedron.

S.G./3. Let T be the set of all lattice points (i.e. all points with integer coordinates) in three-dimensional space. Two such points (x, y, z) and (u, v, w) are called neighbors, if and only if $|x - u| + |y - v| + |z - w| = 1$. Show that there exists a subset S of T such that for each point $p \in T$, there is exactly one point of S among p and its neighbors.

S.G./4. The tetrahedron T is inscribed in a unit sphere with center O. Determine the radius of the sphere S which passes through the centroid of each face of T. Determine also the distance between O and the center of S as a function of the lengths of the edges of T.

Geometric Inequalities†

G.I./1. Let Γ be a unit circle with center O, and let $P_1, P_2, \ldots, P_n$ be points of Γ such that

$$\overrightarrow{OP_1} + \overrightarrow{OP_2} + \cdots + \overrightarrow{OP_n} = \vec{0}.$$

Prove that $\overline{P_1Q} + \overline{P_2Q} + \cdots + \overline{P_nQ} \geqslant n$ for all points Q.

†Solutions of the supplementary solid geometry problems begin on p. 81.
†Solutions of the supplementary geometric inequalities problems begin on p. 83.

G.I./2. A convex quadrilateral is inscribed in a circle of radius 1. Prove that the (positive) difference u between its perimeter and the sum of the lengths of its diagonals satisfies $0 < u < 2$.

G.I./3. Two equilateral triangles are inscribed in a circle with radius r. Let K be the area of the set consisting of all points interior to both triangles. Prove that $2K \geqslant r^2\sqrt{3}$.

G.I./4. For each point P inside a triangle ABC, let D, E, and F be the points of intersection of the lines AP, BP, and CP with the sides opposite to A, B, and C, respectively. Determine P in such a way that the area of triangle DEF is as large as possible.

G.I./5. Prove that in any parallelepiped the sum of the lengths of the edges does not exceed twice the sum of the lengths of the four principal diagonals.

G.I./6. Suppose that 1985 points are given inside a unit cube. Show that one can always choose 32 of them in such a way that every (possibly degenerate) closed polygon with these points as vertices has perimeter less than $8\sqrt{3}$.

G.I./7. A tetrahedron is inscribed in a unit sphere whose center lies in the interior of the tetrahedron. Show that the sum of the edge lengths of the tetrahedron exceeds 6.

Inequalities†

I/1. Find the maximum value of

$$S = \sin^2 \theta_1 + \sin^2 \theta_2 + \cdots + \sin^2 \theta_n$$

subject to the restrictions $0 \leqslant \theta_i \leqslant \pi$, $\theta_1 + \theta_2 + \cdots + \theta_n = \pi$.

I/2. Let $x_n = \sqrt[2]{2 + \sqrt[3]{3 + \sqrt[4]{\ldots + \sqrt[n]{n}}}}$.

Prove: $x_{n+1} - x_n < 1/n!, \qquad n = 2, 3, \ldots$.

†Solutions of the supplementary inequalities problems begin on p. 94.

I/3. Given are the functions $F(x) = ax^2 + bx + c$ and $G(x) = cx^2 + bx + a$, where

$$|F(0)| \leqslant 1, \quad |F(1)| \leqslant 1, \quad \text{and} \quad |F(-1)| \leqslant 1.$$

Prove that, for $|x| \leqslant 1$,   (i) $|F(x)| \leqslant 5/4$;   (ii) $|G(x)| \leqslant 2$.

I/4. Prove that for all $n \geqslant 2$,

$$\frac{x_1^2}{x_1^2 + x_2 x_3} + \frac{x_2^2}{x_2^2 + x_3 x_4} + \cdots + \frac{x_{n-1}^2}{x_{n-1}^2 + x_n x_1} + \frac{x_n^2}{x_n^2 + x_1 x_2} \leqslant n - 1,$$

where all x_i are positive real numbers.

Combinatorics, Probability†

C/1. A $2 \times 2 \times 12$ hole in a wall is to be filled with twenty-four $1 \times 1 \times 2$ bricks. In how many different ways can this be done if the bricks are indistinguishable?

C/2. Let $P_1, P_2, \ldots, P_n$ be distinct two-element subsets of the set of elements $\{a_1, a_2, \ldots, a_n\}$ such that if $P_i \cap P_j \neq \varnothing$, then $\{a_i, a_j\}$ is one of the P's. Prove that each of the a's appears in exactly two of the P's.‡

C/3. Ten airlines serve a total of 1983 cities. There is direct service (without a stopover) between any two cities, and if an airline offers a direct flight from A to B, it also offers a direct flight from B to A. Prove that at least one of the airlines provides a round trip with an odd number of landings.‡

C/4. A box contains p white balls and q black balls, and beside the box lies a large pile of black balls. Two balls chosen at random (with equal likelihood) are taken out of the box. If they are of the same color, a black ball from the pile is put into the box; otherwise, the white ball is put back into the box. The procedure is repeated until the last two balls are removed from the box and one last ball is put in. What is the probability that this last ball is white?

C/5. A fair coin is tossed repeatedly until there is a run of an odd number of heads followed by a tail. Determine the expected number of tosses.

---

†Solutions of the supplementary combinatorics, probability problems begin on p. 99.

‡It is tacitly implied that all elements of any set are distinct; we use the term "multi-set" if its elements need not all be distinct.

Solutions

Twentieth International Olympiad, 1978

1978/1. Since 1978^n and 1978^m agree in their last three digits, the difference

$$1978^n - 1978^m = 1978^m(1978^{n-m} - 1)$$

is divisible by $10^3 = 2^3 \cdot 5^3$; and since the second factor above is odd, 2^3 divides the first. Also

$$1978^m = 2^m \cdot 989^m,$$

so $m \geqslant 3$.

We can write $m + n = (n - m) + 2m$; to minimize this sum, we take $m = 3$ and seek the smallest value of $d = n - m$, such that $1978^d - 1$ is divisible by $5^3 = 125$, i.e.

$$1978^d \equiv 1(\mathrm{mod}\, 125).$$

We shall twice make use of the following

LEMMA. *Let d be the smallest exponent such that $a^d \equiv 1(\mathrm{mod}\, N)$. Then any other exponent g for which $a^g \equiv 1(\mathrm{mod}\, N)$ is a multiple of d.*

PROOF: If d does not divide g, then $g = qd + r$ with $0 < r < d$, and $a^g = a^{qd}a^r \equiv 1(\mathrm{mod}\, N)$ implies $a^r \equiv 1(\mathrm{mod}\, N)$ with $0 < r < d$, contradicting the minimality of d. So $d|g$.

Fermat's theorem states:[†] For any prime p and any integer a not divisible by p,

(1) $$a^{p-1} \equiv 1(\mathrm{mod}\, p).$$

For example,

$$1978^4 \equiv 1(\mathrm{mod}\, 5).$$

[†]For a proof, see e.g. p. 126 of S. L. Greitzer, *The International Mathematical Olympiads*, vol. 27 in this NML series.

Euler extended Fermat's theorem† and showed that

$$a^{\phi(k)} \equiv 1(\bmod k) \quad \text{for } a, k \text{ relatively prime,}$$

where $\phi(k)$ denotes the number of positive integers $\leqslant k$ and relatively prime to k. If $k = p^s$ is a power of a prime, it is easy to see that $\phi(p^s) = p^{s-1}(p-1)$. Therefore, for $a = 1978$ and $p^s = 5^3$, $\phi(125) = 5^2 \cdot 4 = 100$, and

$$1978^{\phi(125)} = 1978^{100} \equiv 1(\bmod 125).$$

By our lemma, d is a divisor of 100.

Since $1978^d - 1$ is divisible by 125, it is certainly divisible by 5, so

$$1978^d \equiv 3^d \equiv 1(\bmod 5).$$

We shall use this last congruence to restrict further our candidates for the exponent d. The smallest positive integer j such that $3^j \equiv 1(\bmod 5)$ is 4. By our lemma, d is a multiple of 4 and a divisor of 100, so d must have one of the three values 4, 20, 100.

We now rule out the first two:

$$(1978)^4 = (2000 - 22)^4 \equiv (-22)^4(\bmod 125)$$

$$\equiv (-2)^4(11)^4 \equiv (4 \cdot 121)^2 \equiv [4(-4)]^2(\bmod 125)$$

$$\equiv (-16)^2 \equiv 256 \equiv 6 \not\equiv 1(\bmod 125).$$

So $d = n - m \neq 4$. We use this computation to find that

$$(1978)^{20} = (1978)^4(1978)^{16} \equiv 6 \cdot 6^4 \equiv 6 \cdot 46 \equiv 26 \not\equiv 1(\bmod 125),$$

so $d = n - m \neq 20$. Therefore $n - m = 100$, $m = 3$, and $n + m = 106$.

1978/2. This problem, stated in three-dimensional space, has an n-dimensional analogue. Its solution is implied by the theorem stated below.

THEOREM. *Let S be a sphere with center O, radius R; let P be a point inside S. Suppose that* $PU_1, PU_2, \ldots, PU_k$ $(k \leqslant n)$
  (i) *are mutually orthogonal and*
  (ii) *the* U_i *are on* S.
Set vectors

$$\mathbf{x}_i = \overrightarrow{PU_i}, \qquad \mathbf{p} = \overrightarrow{OP},$$

and let

  (iii) $$\mathbf{q} = \overrightarrow{OQ} = \mathbf{p} + \mathbf{x}_1 + \mathbf{x}_2 + \cdots + \mathbf{x}_k.$$

†Ibid., p. 136

Then

(1) $$|\mathbf{q}|^2 = kR^2 - (k-1)|\mathbf{p}|^2.$$

Conversely, given a sphere S, a point P inside, and a point Q on the sphere (1), *and a k-dimensional hyperplane H containing P and Q, there exists at least one set of points* $U_1, U_2, \ldots, U_k$ *in H having properties* (*i*), (*ii*) *and* (*iii*) *above.*

In the geometric language of the problem statement, the desired locus of Q is a sphere with center O and radius (1); when $k = 3$, that radius is

$$|\mathbf{q}| = \sqrt{3R^2 - 2|\mathbf{p}|^2}.$$

We shall prove this theorem using vector methods. We shall denote vectors by bold face lower case letters corresponding to the points labelled by capital letters, as we did in stating the theorem. We shall deal with vector sums, as we did in (*iii*), and make frequent use of the dot product $\mathbf{y} \cdot \mathbf{z}$ of two vectors, particularly to write the square of the length of $\mathbf{y}$ as $|\mathbf{y}|^2 = \mathbf{y} \cdot \mathbf{y}$ and to express orthogonality of $\mathbf{y}$ and $\mathbf{z}$ by $\mathbf{y} \cdot \mathbf{z} = 0$. See *Vectors* in the Glossary.

PROOF: We have $\overrightarrow{PU_i} = \mathbf{u}_i - \mathbf{p} = \mathbf{x}_i$ and $\mathbf{x}_i \cdot \mathbf{x}_j = 0$ for $i \neq j$ by property (*i*). We use this and (*iii*) to compute

$$|\mathbf{q}|^2 = \left(\mathbf{p} + \sum_1^k \mathbf{x}_i\right) \cdot \left(\mathbf{p} + \sum_1^k \mathbf{x}_i\right) = |\mathbf{p}|^2 + 2\mathbf{p} \cdot \sum_1^k \mathbf{x}_i + \sum_1^k |\mathbf{x}_i|^2$$

$$= |\mathbf{p}|^2 + 2\mathbf{p} \cdot \sum_1^k (\mathbf{u}_i - \mathbf{p}) + \sum_1^k |\mathbf{u}_i - \mathbf{p}|^2$$

$$= |\mathbf{p}|^2 + 2\mathbf{p} \cdot \sum_1^k \mathbf{u}_i - 2k|\mathbf{p}|^2 + \sum_1^k |\mathbf{u}_i|^2 - 2\mathbf{p} \cdot \sum_1^k \mathbf{u}_i + k|\mathbf{p}|^2.$$

According to Property (*ii*), $|\mathbf{u}_i|^2 = R^2$, so

$$|\mathbf{q}|^2 = kR^2 - (k-1)|\mathbf{p}|^2.$$

We prove the converse by induction on k. When $k = 1$, $|\mathbf{q}|^2 = R^2$; H is the line through P and Q, and $U_1 = P$ trivially satisfies the requirements.

For $k > 1$, let T be the sphere with diameter $\overrightarrow{PQ}$. By Thales' theorem, T is the set of all points U such that $\overrightarrow{PU} \perp \overrightarrow{UQ}$, i.e.

(2) $$(\mathbf{u} - \mathbf{p}) \cdot (\mathbf{q} - \mathbf{u}) = 0.$$

Since P and Q are on T and $|\mathbf{p}| < R < |\mathbf{q}|$, T intersects S.

Let U_k be any point on $T \cap S$, and define $\mathbf{x}_k = \mathbf{u}_k - \mathbf{p}$. Then the point V such that

(3) $$\mathbf{v} = \mathbf{q} - \mathbf{x}_k = \mathbf{p} + \mathbf{q} - \mathbf{u}_k$$

is the fourth vertex of rectangle PU_kQV. [This follows from the parallelogram law of addition and the fact that $PU_k \perp QU_k$.] We use (3) to compute

$$|\mathbf{v}|^2 = |\mathbf{p} + \mathbf{q}|^2 - 2\mathbf{u}_k \cdot (\mathbf{p} + \mathbf{q}) + |\mathbf{u}_k|^2$$
$$= |\mathbf{p}|^2 + 2\mathbf{p} \cdot \mathbf{q} + |\mathbf{q}|^2 - 2\mathbf{u}_k \cdot (\mathbf{p} + \mathbf{q}) + |\mathbf{u}_k|^2,$$

and we use (2) to eliminate $2[\mathbf{p} \cdot \mathbf{q} - \mathbf{u}_k \cdot (\mathbf{p} + \mathbf{q})]$:

$$(\mathbf{u}_k - \mathbf{p}) \cdot (\mathbf{u}_k - \mathbf{q}) = |\mathbf{u}_k|^2 - \mathbf{u}_k \cdot (\mathbf{p} + \mathbf{q}) + \mathbf{p} \cdot \mathbf{q} = 0,$$
$$\mathbf{p} \cdot \mathbf{q} - \mathbf{u}_k \cdot (\mathbf{p} + \mathbf{q}) = -|\mathbf{u}|^2.$$

So by our hypothesis (1), we obtain

$$|\mathbf{v}|^2 = |\mathbf{p}|^2 + |\mathbf{q}|^2 + |\mathbf{u}_k|^2 - 2|\mathbf{u}_k|^2$$
$$= |\mathbf{q}|^2 - R^2 - |\mathbf{p}|^2$$
$$= (k - 1)R^2 - (k - 2)|\mathbf{p}|^2.$$

Alternatively, we could have determined $|\mathbf{v}|^2$ by applying the following elementary and useful

LEMMA. *If $ABCD$ is a rectangle with diagonals AC and BD, and if O is any point, then*

$$OA^2 + OC^2 = OB^2 + OD^2.$$

PROOF: Choosing a vector origin at the center of the rectangle, the vertices and the point O (which need not be in the plane of the rectangle) have a vector representation $\mathbf{m}$, $\mathbf{n}$, $-\mathbf{m}$, $-\mathbf{n}$ and $\mathbf{o}$, respectively, where $|\mathbf{m}| = |\mathbf{n}|$. It then follows that

$$|\mathbf{o} + \mathbf{m}|^2 + |\mathbf{o} - \mathbf{m}|^2 = |\mathbf{o} + \mathbf{n}|^2 + |\mathbf{o} - \mathbf{n}|^2.$$

When we apply this lemma to rectangle PU_kQU, we find

$$|\mathbf{v}|^2 + |\mathbf{u}_k|^2 = |\mathbf{p}|^2 + |\mathbf{q}|^2, \quad \text{so } |\mathbf{v}|^2 = \mathbf{q}^2 - R^2 + |\mathbf{p}|^2.$$

When we add $|\mathbf{p}|^2 - R^2$ to $|\mathbf{q}|^2 = kR^2 - (k - 1)|\mathbf{p}|^2$, we get

$$|\mathbf{v}|^2 = (k - 1)R^2 - (k - 2)|\mathbf{p}|^2.$$

Now let $H_\perp$ be the set of points E in H such that $\overrightarrow{PE} \perp \mathbf{x}_k$. Then V and P are in $H_\perp$, and by the induction hypothesis, there are points U_i, $i = 1, 2, \ldots, k - 1$ on S such that the $\mathbf{x}_i = \mathbf{u}_i - \mathbf{p}$ are orthogonal, and $\mathbf{v} = \mathbf{p} + \mathbf{x}_1 + \mathbf{x}_2 + \cdots + \mathbf{x}_{k-1}$.

1978/3 First solution. Let $F = \{f(n)\}$ and $G = \{g(n)\}$, for $n = 1, 2, 3 \ldots$. Now $g(1) = f(f(1)) + 1 > 1$, and hence $f(1) = 1$ and $g(1) = 2$. Postponing further numerical calculations, we prove that if $f(n) = k$, then

(1) $f(k) = k + n - 1,$ (2) $g(n) = k + n,$ and
(3) $f(k + 1) = k + n + 1.$

Taking these for granted for a moment, we apply them to $f(1) = 1$ to get $f(1) = 1$, $g(1) = 2$, and $f(2) = 3$. If we apply equation (1) repeatedly to $f(2) = 3$ and its successors, we get a chain of results

$$f(3) = 4, \quad f(4) = 6, \quad f(6) = 9, \quad f(9) = 14,$$

$$f(14) = 22, \quad f(22) = 35, \quad f(35) = 56,$$

$$f(56) = 90, \quad f(90) = 145, \quad f(145) = 234, \quad f(234) = 378, \ldots .$$

But $f(240)$ is not in this chain. Note that equation (3) produces larger numbers; e.g., applied to $f(145) = 234$ it gives $f(235) = 380$. Moving back a little more in the chain, we find after a trial or two that if we apply equation (3) to $f(56) = 90$ and its successors we get $f(91) = 147$, $f(148) = 239$, and finally $f(240) = 388$.

But this answer to the problem must be justified by proving equations (1), (2), and (3). Assuming that $f(n) = k$, we note that the elements in the two disjoint sets $\{f(1), f(2), f(3), \ldots, f(k)\}$ and $\{g(1), g(2), g(3), \ldots, g(n)\}$ comprise all the natural numbers from 1 to $g(n)$, because $g(n) = f(f(n)) + 1 = f(k) + 1$. Counting elements in sets we see that $g(n) = k + n$ or $g(n) = f(n) + n$. This is equation (2), and equation (1) follows from $k + n = g(n) = f(k) + 1$.

From the equation $g(n) - 1 = f(f(n))$, we note that $g(n) - 1$ is a member of F, and hence no two consecutive integers are members of G. Since $k + n$ is a member of G, it follows that both $k + n - 1$ and $k + n + 1$ are members of F, in fact consecutive members of F. Hence equation (1) implies that $k + n + 1 = f(k + 1)$.

Second solution. Sequences $\{f(k)\}, \{g(k)\}$ of integers which, together, contain every integer exactly once are called *complementary*. Special sets of complementary sequences were constructed by S. Beatty. He showed that for any pair (α, β) of positive irrational numbers satisfying

(4)
$$\frac{1}{\alpha} + \frac{1}{\beta} = 1,$$

the sequences

$$F(n) = \{[n\alpha]\}, \quad G(n) = \{[n\beta]\}, \qquad n = 1, 2, \ldots$$

(where $[z]$ denotes the integer part of z) are complementary.†

---

†A short, simple proof of Beatty's theorem is given on p. 93 of R. Honsberger's book *Ingenuity in Mathematics*, vol. 23 in this NML series, in an essay entitled "Complementary Sequences.". The essay discusses other interesting properties of such sequences and gives references for further study. See also Problem #3177 by S. Beatty, Amer. Math. Monthly, 34 (1927) 159. See also S. W. Golomb, *The "Sales Tax" Theorem*, Math. Mag. 49 (1976) p. 187 and H. Grossman, *A Set Containing All Integers*, Am. Math. Monthly (1962) p. 532.

We claim that our sequences $\{f(n)\}$, $\{g(n)\}$ are Beatty sequences. The key to the identification lies in relation (2): $g(n) = f(n) + n$, which translates into $[n\beta] = [n\alpha] + n = [n\alpha + n]$. This holds for all n if $n\beta = n\alpha + n$, i.e. if

(5) $\beta = \alpha + 1.$

Substituting this for β in (4), we find that $\dfrac{1}{\alpha} + \dfrac{1}{\alpha + 1} = 1$, or equivalently, that

$$\alpha^2 - \alpha - 1 = 0.$$

We choose the positive root $\alpha = \frac{1}{2}(1 + \sqrt{5})$, so $\beta = \frac{1}{2}(3 + \sqrt{5})$, and compute

$$f(240) = [240\alpha] = \left[120(1 + \sqrt{5})\right] = 120 + [120\sqrt{5}].$$

Now $\sqrt{5} \approx 2.236$, $120\sqrt{5} \approx 12(22.36) = 264 + 4.32$, so $[120\sqrt{5}] = 268$ and

$$f(240) = 120 + 268 = 388.$$

Note. A third, very instructive solution will only be sketched here. Let $b_0 = 1$, $b_1 = 2$, $b_2 = 3$, $b_3 = 5$, $b_4 = 8, \ldots, b_i = F_{i+2} = (i + 2)$nd Fibonacci number. Then every positive integer n has a unique representation of the form

$$n = a_k b_k + a_{k-1} b_{k-1} + \cdots + a_0 b_0,$$

where $a_i = 0$ or 1 and no two consecutive a_i are 1. Then $a_k a_{k-1} \ldots a_0$ is the representation of n in the "Fibonacci base." For example $1 = 1$, $2 = 10$, $3 = 100$, $4 = 101$, $5 = 1000$, $6 = 1001$, $7 = 1010$, etc. The largest d-digit number, $F_{d+2} - 1$, is $10101\ldots$.

The sequence of numbers which *end in an even number of zeros* (possibly none) in the Fibonacci base and the numbers which *end in an odd number of zeros* form *complementary sequences*. Let us call the n-th member of the first sequence $f(n)$, and the n-th member of the second sequence $g(n)$. We present a way of computing $f(n)$ in the Fibonacci base, and it will be easy to check that $g(n) = f(f(n)) + 1$, and hence that these two sequences are the ones in the given problem.

Let $n = a_k a_{k-1} \ldots a_0$ in the Fibonacci base. We claim $f(n)$ is obtained as follows: Form $a_k a_{k-1} \ldots a_0 0$. If this ends with an even number of 0's, it is $f(n)$. If it ends with an odd number of 0's, $\ldots 100 \ldots 0$, then subtract 1, which in the Fibonacci base means replacing the tail segment $\ldots 1000 \ldots 0$ by the equally long segment $\ldots 0101 \ldots 01$. It is easy to see that this process gives all numbers ending with an even number of 0's in increasing order, hence it computes $f(n)$. If $f(n) = c_k c_{k-1} \ldots c_0$ ends with an even number of 0's, then by the construction above $f(f(n)) + 1$ is $c_k c_{k-1} \ldots c_0 0$. Thus $f(f(n)) + 1$ assumes all values ending in an odd number of 0's, and hence it is $g(n)$.

1978/4. Since $AB = AC$, our figure is symmetric with respect to diameter AM, where M is the point of tangency of the two circles. AM

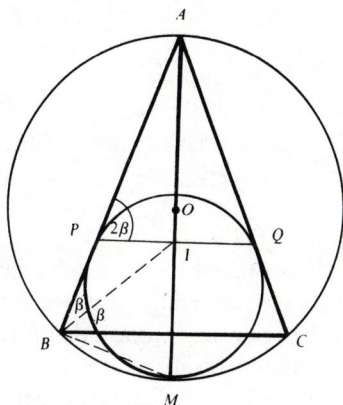

bisects $\angle A$, $\angle PMQ$ and segment PQ, which is parallel to BC and whose midpoint we denote by I. Letting $\angle APQ = 2\beta$, we have also $\angle ABC = 2\beta$. Moreover, $\angle PMQ = \angle APQ = 2\beta$ since both are measured by half of arc PQ. Thus $\angle PMI = \frac{1}{2}\angle PMQ = \beta$.

Since $\angle ABM$ and $\angle MIP$ are right angles, $BMIP$ can be inscribed in a circle in which angles PBI and PMI subtend the same arc. Hence $\angle PBI = \angle PMI = \beta$. Thus the angle bisectors from A and B in $\triangle ABC$ intersect in I, so I is the incenter of $\triangle ABC$.

1978/5. If we had a monotonic sequence $\{m_i\}$ of positive integers, i.e. $m_i < m_j$ for $i < j$, then for each k, $m_k \geqslant k$. Hence $m_k/k^2 \geqslant 1/k$, and so

$$\sum_1^n \frac{m_k}{k^2} \geqslant \sum_1^n \frac{1}{k} \qquad \text{for every } n.$$

But the $\{a_i\}$ in the problem is not necessarily monotonic. Actually, this works in our favor, as we shall show by proving the following more general rearrangement inequalities [107, p. 261]:

Let $\{r_i\}$ and $\{s_i\}$, $i = 1, 2, \ldots, n$, be two sets of real numbers such that

$$r_1 \leqslant r_2 \leqslant \cdots \leqslant r_n \qquad \text{and} \qquad s_1 \leqslant s_2 \leqslant \cdots \leqslant s_n.$$

Then, if $\{t_i\}$ is a permutation of the s_i,

$$(1) \qquad \sum_{i=1}^n r_i s_i \geqslant \sum_{i=1}^n r_i t_i \geqslant r_1 s_n + r_2 s_{n-1} + \cdots + r_n s_1.$$

PROOF: Suppose for some j, k with $j < k$, we have $t_j \geqslant t_k$. Then

$$r_j t_k + r_k t_j - (r_j t_j + r_k t_k) = (t_j - t_k)(r_k - r_j) \geqslant 0.$$

This implies

$$r_j t_k + r_k t_j \geq r_j t_j + r_k t_k,$$

and we do not decrease $\sum_{i=1}^{n} r_i t_i$ by switching t_j with t_k. A finite number of such exchanges leads to an increasing order of the t_i, so that

$$\sum_{i=1}^{n} r_i s_i \geq \sum_{i=1}^{n} r_i t_i.$$

The right inequality in (1) is proved by the same argument.

The inequalities (1) become physically intuitive if we interpret the r_i as distance positions from the fulcrum on one side of a see-saw and the t_i as the weight of a person sitting at position r_i. The greatest turning moment about the fulcrum is obtained by the heaviest person sitting at the end position and so on.

To solve our problem, take T to be the first n members of the given sequence $\{a_k\}$ of distinct positive integers, and let R be the set of n numbers $1/k^2$, $k = 1, 2, \ldots, n$. Then

$$(2) \quad \sum_{k=1}^{n} \frac{a_k}{k^2} \geq (\text{smallest } a) \cdot \frac{1}{1^2} + (\text{second smallest } a) \cdot \frac{1}{2^2} + \cdots + (\text{largest } a) \frac{1}{n^2}.$$

Since the smallest a is ≥ 1, the a next in size is $\geq 2, \ldots$ and the largest a is $\geq n$, the sum on the right in (2) is greater than or equal to

$$1 + \frac{2}{2^2} + \frac{3}{3^2} + \cdots + \frac{n}{n^2} = \sum_{k=1}^{n} \frac{1}{k}.$$

1978/6. The problem may be stated as follows: If the set $\{1, 2, 3, \ldots, 1978\}$ is separated in any way into six disjoint subsets A, B, C, D, E and F, then at least one of these six sets contains the sum of two (possibly equal) elements of that set.

First we observe that no matter how 1978 elements are distributed among 6 sets, at least one of these sets, say A, contains at least $\lceil 1978/6 \rceil = 330$ members,† which we denote by a_i and order so that $a_1 < a_2 < \cdots < a_{330}$. If any of the 329 differences

$$b_1 = a_{330} - a_{329}, \quad b_2 = a_{330} - a_{328}, \ldots, \quad b_{329} = a_{330} - a_1,$$

belongs to A, the problem is solved; because then for some k,

$$a_{330} - a_k = a_l \in A, \quad \text{so} \quad a_k + a_l = a_{330}.$$

So suppose none of the 329 differences $a_{330} - a_i$ belongs to A; instead they are distributed among the five remaining sets. Again we conclude that

†$\lceil x \rceil$ denotes the least integer $\geq x$.

at least one of these, say B, contains at least $\lceil 329/5 \rceil = 66$ of these differences. We take *their* differences $b'_{66} - b'_{66-j} = c_j$.

We describe this process in more detail: We order the differences $b_j = a_{330} - a_{330-j}$ which are in B according to increasing size and rename them $b'_1, b'_2, \ldots, b'_{66}$. We form

$$c_j = b'_{66} - b'_{66-j}, \qquad\qquad j = 1, 2, \ldots 65$$

$$= (a_{330} - a_i) - (a_{330} - a_k), \qquad i < k < 330$$

$$= a_k - a_i, \qquad\qquad 1 \leqslant a_i < a_k \leqslant 1978.$$

Thus the c_j are integers satisfying $1 < c_j < 1977$.

If any of these c_j occur in either A or B, the problem is solved. If not, they are distributed among the four remaining sets at least one of which, say C, contains at least $\lceil 65/4 \rceil = 17$ of them. Again we order their 16 differences $c_{17} - c_j$ according to size, rename then c'_i so that $c'_i < c'_j$ for $i < j$. If none are elements of sets A, B or C, all are distributed among the three remaining sets, at least one of which, say D, contains at least $\lceil 17/3 \rceil = 6$ elements, d'_i.

Similarly, we argue: if their 5 differences are not in A, B, C, or D, at least one of the two remaining sets, say E, contains at least $\lceil 5/2 \rceil = 3$ of them, and if neither of their two differences e'_1, e'_2 is in one of the sets A, B, C, D, E, they must both be in F. Their difference $e'_2 - e'_1 < 1978$ must belong to one of the six sets.

Twenty-first International Olympiad, 1979

1979/1. The negative terms have even denominators; we write each $-1/2k$ as

$$\frac{1}{2k} - \frac{1}{k}.$$

So

$$\frac{p}{q} = \left(1 + \frac{1}{2} + \frac{1}{3} + \cdots + \frac{1}{1319}\right) - 2\left(\frac{1}{2} + \frac{1}{4} + \frac{1}{6} + \cdots + \frac{1}{1318}\right)$$

$$= \left(1 + \frac{1}{2} + \frac{1}{3} + \cdots + \frac{1}{1319}\right) - \left(1 + \frac{1}{2} + \frac{1}{3} + \cdots + \frac{1}{659}\right)$$

$$= \frac{1}{660} + \frac{1}{661} + \cdots + \frac{1}{1319}.$$

Since

$$\frac{1}{660 + j} + \frac{1}{1319 - j} = \frac{1319 + 660}{(660 + j)(660 - j)}$$

for all j, we get a constant numerator when we add the fractions in pairs:

$$\frac{p}{q} = \left(\frac{1}{660} + \frac{1}{1319}\right) + \left(\frac{1}{661} + \frac{1}{1318}\right) + \cdots + \left(\frac{1}{989} + \frac{1}{990}\right)$$

$$= \frac{1979}{660 \cdot 1319} + \frac{1979}{661 \cdot 1318} + \cdots + \frac{1979}{989 \cdot 990}$$

$$= 1979\frac{p'}{q'},$$

where q' is the product of all the integers from 660 to 1319, each relatively prime to 1979 (which is a prime). Now

$$pq' = 1979p'q;$$

and since 1979 does not divide q', it must divide p.

1979/2. We first show that the edges A_1A_2, A_2A_3, A_3A_4, A_4A_5, A_5A_1 are all the same color by indirect proof. Suppose on the contrary that, say, edge A_1A_2 is red and A_2A_3 is green. At least three of the five segments A_2B_1, A_2B_2, A_2B_3, A_2B_4, A_2B_5 have the same color. Suppose without loss of generality that these are red, and label them A_2B_i, A_2B_j, A_2B_k. Then at least one of the segments B_iB_j, B_jB_k, B_kB is an edge of the base; call it B_rB_s. If B_rB_s were red, we would have a red triangle $A_2B_rB_s$. Therefore B_rB_s is green. Now segments A_1B_r and A_1B_s must also be green, for otherwise we would have $A_1A_2B_r$ or $A_1A_2B_s$ as red triangles. Therefore $A_1B_rB_s$ is a green triangle. This contradiction implies that A_1A_2 and A_2A_3 have the same color and similarly that all the edges of each base have the same color.

Now suppose the top edges are all red and the bottom edges are all green. If 3 green edges join A_1 to the bottom, 2 of them must terminate on adjacent vertices B_r, B_s of the base. Then $A_1B_rB_s$ is a green triangle, a contradiction. Hence at least 3 red edges join A_1 to the bottom. Similarly, at least 3 red edges join A_2 to the bottom. Since we now have 6 red edges, at least 2 of them must terminate on the same vertex B_i of the bottom. Then $A_1A_2B_i$ is a red triangle, a contradiction.

Remark. The same argument works if the top and bottom faces are polygons with $2n + 1$ sides. However, if the top and bottom are polygons with $2n$ sides, the conclusion is false. For instance, a counterexample is obtained by painting the top edges red, the bottom edges green, and the edge A_iB_j red or green according as $i - j$ is even or odd.

1979/3. Both moving points have the same angular speed (since they both return to A after one revolution). In the figure below, point Q_1 on circle C_1 and point Q_2 on circle C_2 have both travelled counterclockwise an arbitrary angle θ. Also, B is the other point of intersection of the two circles, and MAN is a line segment perpendicular to AB. Since $\angle ABQ_1 = \theta/2$ and $\angle ABQ_2 = \pi - \theta/2$, the line Q_1Q_2 passes through B for all θ. Since $\angle MAB = \angle NAB = 90°$, MB and NB are diameters; hence $\angle MQ_1B = \angle BQ_2N = 90°$. It now follows that MQ_1 and NQ_2 are parallel. Thus the perpendicular bisector of Q_1Q_2 (i.e. the set of points equidistant from Q_1 and Q_2) intersects the fixed segment MN in its midpoint, P, for all positions Q_1, Q_2 of the moving points.

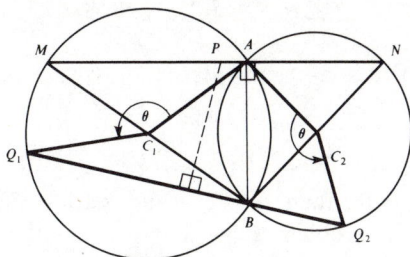

1979/4. For any point R (to be determined), denote the line through R and P by l, and set $\angle QPR = 2\theta$. We now locate point S on l such that $QP = SP$ as shown in the figure. Then $SR = QP + PR$. By the Law of Sines,

$$\lambda \equiv \frac{QP + PR}{QR} = \frac{SR}{QR} = \frac{\sin \angle SQR}{\sin \theta}.$$

For all points R on l, λ will be maximized when $\angle SQR = 90°$. It now remains to maximize $1/\sin \theta$. We achieve this when 2θ is a minimum, and this occurs when l passes through P and the foot T of the perpendicular from Q to plane π.

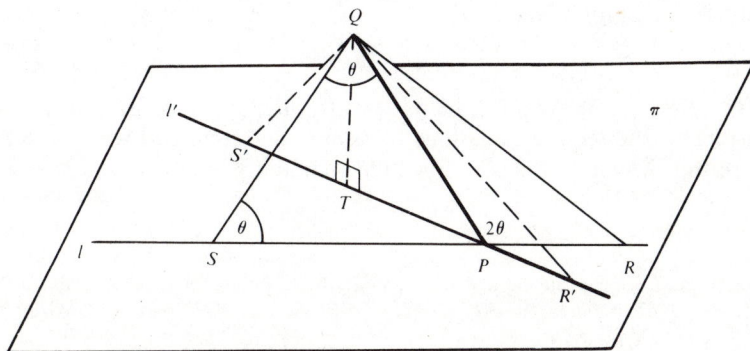

If T and P are distinct points, R is then uniquely determined. If not, R can be any point on a circle of radius QP centered at P.

1979/5 First Solution. The required relations imply

$$a^2 \sum_1^5 k x_k + \sum_1^5 k^5 x_k = 2a \sum_1^5 k^3 x_k$$

or equivalently

$$\sum_1^5 k(a - k^2)^2 x_k = 0.$$

Hence, each of the five non-negative terms must vanish:

$$k(a - k^2)^2 x_k = 0, \qquad\qquad k = 1, 2, 3, 4, 5.$$

Thus either $x_1 = x_2 = x_3 = x_4 = x_5 = 0$, or else four of the x_k are zero. If for some j, $x_j \neq 0$, then $a = j^2$, and setting this into any of the relations stated in the problem gives $x_j = j$. Thus the six possible values for a are

$$0, \quad 1, \quad 2^2 = 4, \quad 3^2 = 9, \quad 4^2 = 16, \quad \text{and} \quad 5^2 = 25.$$

Second solution. The Cauchy inequality† for real numbers b_i, c_i states that

(1)
$$\left(\sum b_k c_k \right)^2 \leqslant \left(\sum b_k^2 \right) \left(\sum c_k^2 \right)$$

where equality holds if and only if $c_k = \lambda b_k$, $k = 1, 2, 3, 4, 5$. Set

$$b_k = \sqrt{k x_k}, \quad c_k = \sqrt{k^5 x_k}; \quad \text{then } b_k c_k = k^3 x_k,$$

and (1) reads

$$\left(\sum k^3 x_k \right)^2 = (a^2)^2 \leqslant \left(\sum k x_k \right)\left(\sum k^5 x_k \right) = a^4$$

Therefore equality holds in (1), so $c_k = \lambda b_k$, or $k^5 x_k = \lambda^2 k x_k$, i.e.

(2)
$$x_k(k^4 - \lambda^2) = 0, \qquad\qquad k = 1, 2, 3, 4, 5.$$

If, for some j, $x_j \neq 0$, (2) yields $\lambda = j^2$, and $x_k = 0$ for $k \neq j$. Again the sums in the required relations have only one term and yield $a = j^2$ as before. So $\lambda = a$; we again conclude that the possible values for a are $0, 1, 4, 9, 16, 25$.

---

†For a proof, see e.g. *Introduction to Inequalities* by E. Beckenbach and R. Bellman, NML vol. 3, pp. 62–67 or *International Mathematical Olympiads 1959–77* by S. L. Greitzer, NML vol. 27, p. 128. Also see Glossary.

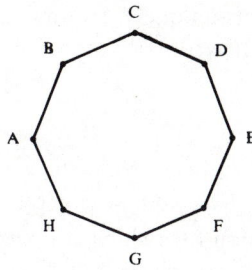

1979/6 First solution. The frog cannot reach E in less than 4 jumps, so $a_1 = a_2 = a_3 = 0$; he can reach E in 4 jumps going in either direction, so

$$a_4 = 2.$$

An odd number of jumps from A will get the frog to B, D, F, H, but never to E (see figure); so

$$a_{2n-1} = 0 \qquad \text{for all } n.$$

Now let b_n be the number of distinct paths of exactly n jumps starting at C and ending at E (this is the same as the number starting at G). There is just one way to get from C to E in two jumps, so

$$b_2 = 1.$$

After the first 2 jumps from A, the frog is either at C, or at G, or back at A. He can get back to A in 2 ways: $A \to B \to A$ or $A \to H \to A$. So the number a_n of paths from A to E is made up of the number of paths of $n - 2$ jumps from C, from G, and twice the number from A. Thus

(1) $$a_n = 2b_{n-2} + 2a_{n-2}.$$

When the frog starts at C and makes more than two jumps, the first two either leave him at A or back at C (since his landing at E would mean he made exactly 2 jumps). So the number b_n, $n > 2$, of paths from C to E is made of the number of paths of $n - 2$ jumps from A and twice the number of paths of $n - 2$ jumps from C (since the first 2 jumps could have been $C \to D \to C$ or $C \to B \to C$); so

(2) $$b_n = 2b_{n-2} + a_{n-2}, \qquad n > 2.$$

Subtracting (1) from (2), we obtain

$$b_n - a_n = -a_{n-2},$$

and replacing n with $n - 2$, we have

$$b_{n-2} = a_{n-2} - a_{n-4}$$

which, when substituted into (1) gives

(3) $$a_n = 4a_{n-2} - 2a_{n-4} \qquad \text{for } n > 4.$$

Clearly the sequence a_n is uniquely determined by $a_2 = 0$, $a_4 = 2$ and relation (3) for $n > 4$. Therefore we need only verify that the alleged solution

$$c_{2n} = \frac{1}{\sqrt{2}}(x^{n-1} - y^{n-1})$$

given in the statement of the problem satisfies $c_2 = 0$ and $c_4 = 2$, the initial values for the recursion (3) (which it does); and also that it satisfies (3) with n replaced by $2n$, i.e. $c_{2n} = 4c_{2n-2} - 2c_{2n-4}$.

Since $x = 2 + \sqrt{2}$ and $y = 2 - \sqrt{2}$ are roots of

$$t^2 - 4t + 2 = 0,$$

they are roots also of

$$p(t) = t^{n-1} - 4t^{n-2} + 2t^{n-3} = 0.$$

Thus $(1/\sqrt{2})(p(x) - p(y)) = 0$, or

$$\frac{x^{n-1} - y^{n-1}}{2} - 4\frac{x^{n-2} - y^{n-2}}{2} + 2\frac{x^{n-3} - y^{n-3}}{2} = 0,$$

and

$$c_{2n} = 4c_{2n-2} - 2c_{2n-4}.$$

By the uniqueness of a solution of (3) with given initial conditions, we may now identify a_{2n} with c_{2n};

$$a_{2n} = c_{2n} = \frac{1}{\sqrt{2}}(x^{n-1} - y^{n-1}).$$

In this solution, we used the given formula and verified that it satisfies the recursion relation (3) we derived. But suppose no formula had been given; could we have found it? The answer is yes, and we substantiate it below:

Second solution. We derive equations (1) and (2) as in the first solution. But instead of eliminating the number b_n of frog paths from C, we work with the pair a_n, b_n simultaneously.

We observed earlier that $a_2 = 0$, $b_2 = 1$. Since the frog cannot reach E from either A or C in an odd number of jumps, we set $n = 2m$ and think of the system

(1) $a_n = 2a_{n-2} + 2b_{n-2}$

(2) $b_n = a_{n-2} + 2b_{n-2}$

as a transformation of a vector $\mathbf{v}_{m-1}$ into $\mathbf{v}_m = \begin{pmatrix} a_{2m} \\ b_{2m} \end{pmatrix}$ by means of the matrix

$$T = \begin{pmatrix} 2 & 2 \\ 1 & 2 \end{pmatrix},$$

so that

$$\mathbf{v}_m = T\mathbf{v}_{m-1}, \qquad \mathbf{v}_1 = \begin{pmatrix} a_2 \\ b_2 \end{pmatrix} = \begin{pmatrix} 0 \\ 1 \end{pmatrix}.$$

The eigenvalues† of T are the solutions λ_1, λ_2 of the characteristic equation† for T:

$$\begin{vmatrix} 2 - \lambda & 2 \\ 1 & 2 - \lambda \end{vmatrix} = \lambda^2 - 4\lambda + 2 = 0;$$

they are $\lambda_1 = 2 + \sqrt{2}$, $\lambda_2 = 2 - \sqrt{2}$. The corresponding eigenvectors† $\mathbf{u}_1, \mathbf{u}_2$ have the property

(4) $$T\mathbf{u}_i = \lambda_i\mathbf{u}, \quad T[T\mathbf{u}_i] = \lambda_i^2\mathbf{u}_i, \ldots, \quad T^{(n)}\mathbf{u}_i = \lambda_i^n\mathbf{u}_i$$

for $i = 1, 2$. We find them from (4), i.e. by solving

$$\begin{pmatrix} 2 & 2 \\ 1 & 2 \end{pmatrix}\begin{pmatrix} \alpha_i \\ \beta_i \end{pmatrix} = \lambda_i\begin{pmatrix} \alpha_i \\ \beta_i \end{pmatrix},$$

and find that

$$\alpha_1/\beta_1 = \sqrt{2}, \qquad \alpha_2/\beta_2 = -\sqrt{2}.$$

We set

$$\mathbf{u}_1 = \frac{1}{\sqrt{2}}\begin{pmatrix} 1 \\ 1/\sqrt{2} \end{pmatrix}, \qquad \mathbf{u}_2 = \frac{1}{\sqrt{2}}\begin{pmatrix} -1 \\ 1/\sqrt{2} \end{pmatrix}.$$

Then $\mathbf{v}_1 = \begin{pmatrix} 0 \\ 1 \end{pmatrix}$ is the linear combination

$$\mathbf{v}_1 = \lambda_1\mathbf{u}_1 + \lambda_2\mathbf{u}_2$$

of $\mathbf{u}_1$ and $\mathbf{u}_2$, and

$$\mathbf{v}_m = T^{(m-1)}\mathbf{v}_1 = \lambda_1^{m-1}\mathbf{u}_1 + \lambda_2^{m-2}\mathbf{u}_2 = \begin{pmatrix} a_{2m} \\ b_{2m} \end{pmatrix}.$$

When we write the equation for the first component of $\mathbf{v}_m$, we obtain

$$a_{2m} = \frac{1}{\sqrt{2}}\left((2 + \sqrt{2})^{m-1} - (2 - \sqrt{2})^{m-1}\right).$$

Note. The number a_{2m} is an integer for all m. Since $\lambda_2 = 2 - \sqrt{2} < .6 < 1$, the second term, $\lambda_2^{m-1}/\sqrt{2}$, in a_{2m} is less than $\frac{1}{2}$ even for $m = 2$ and becomes negligibly small as m increases. Thus for all m, a_{2m} is the integer part of the first term: $a_{2m} = [\lambda_1^{m-1}/\sqrt{2}]$.

†For definitions and a more detailed treatment, see 2nd solution of IMO 63/4 in NML vol. 27, p. 61.

Twenty-second International Olympiad, 1981

1981 / 1. Denote the lengths of the sides opposite A, B, C by a, b, c respectively, and those of segments PD, PE, PF by x, y and z (see figure).

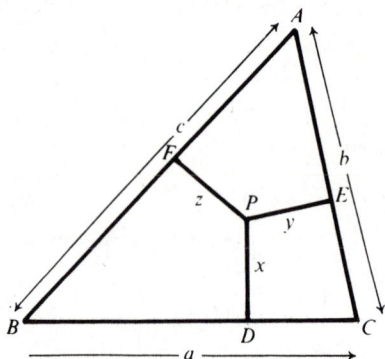

The area K of the triangle satisfies

(1) $$2K = ax + by + cz.$$

We wish to minimize

(2) $$\frac{a}{x} + \frac{b}{y} + \frac{c}{z}$$

subject to constraint (1). We can do this in several ways. The simplest perhaps is to use Cauchy's inequality†

$$(u_1v_1 + u_2v_2 + u_3v_3)^2 \le (u_1^2 + u_2^2 + u_3^2)(v_1^2 + v_2^2 + v_3^2)$$

with $\sqrt{ax}, \sqrt{by}, \sqrt{cz}$ serving as u's and $\sqrt{a/x}, \sqrt{b/y}, \sqrt{c/z}$ as v's. Thus,

(3) $$(a + b + c)^2 \le (ax + by + cz)\left(\frac{a}{x} + \frac{b}{y} + \frac{c}{z}\right) = 2K\left(\frac{a}{x} + \frac{b}{y} + \frac{c}{z}\right),$$

or

$$\frac{a}{x} + \frac{b}{y} + \frac{c}{z} \ge \frac{(a + b + c)^2}{2K}.$$

There is equality if and only if the triples $(a/x, b/y, c/z)$ and (ax, by, cz) are proportional, i.e. if and only if $x = y = z$. Thus the minimum value of (2) occurs when P is the incenter of $\triangle ABC$.

†See Glossary.

The result can be extended to three dimensions for tetrahedra, and to n-dimensional Euclidean space for simplexes (n-dimensional "triangles") in several ways. These are included in the following extremum problem:

Determine the extreme values of $S = \Sigma a_i x_i^p$ given that $\Sigma x_i = 1$. Here the sums are from $i = 1, 2, \ldots, n$; $x_i \geqslant 0$, and the a_i and p are given numbers with $a_i > 0$.

One simple method here is to use Hölder's inequality† which is a generalization of Cauchy's inequality.

Readers acquainted with calculus can use Lagrange's method for optimizing a function $f(x, y, z)$ subject to a constraint $g(x, y, z) = $ constant. However, in the solution it must be verified that a maximum or minimum is actually achieved. For the three variable case of the given problem, this is not hard. However, for the n-variable case, one must also examine *all* the lower dimensional "faces" of the constraint conditions and this is onerous. While the calculus method is a very general one, a solution by means of an appropriate known inequality will be simpler.

1981 / 2 First solution. We partition the collection of subsets containing r elements according to their minimum element k, where $k = 1, 2, \ldots, n - r + 1$. For a minimum element k, the remaining $r - 1$ elements are chosen from the $n - k$ elements $k + 1, k + 2, \ldots, n$. Thus there are $\binom{n - k}{r - 1}$ subsets of r elements which have k as their least element. Therefore

$$F(n, r) = \frac{N(n, r)}{\binom{n}{r}},$$

where

$$(1) \qquad N(n, r) = \sum_{k=1}^{n-r+1} k \binom{n - k}{r - 1}$$

is the sum of all their least members. To find $N(n, r)$, we use the "fundamental summation formula"‡

$$(2) \qquad \sum_{k=1}^{n} (F(k) - F(k - 1)) = F(n) - F(0)$$

twice. We use the familiar binomial coefficient identity

$$\binom{n - k}{r - 1} = \binom{n - k + 1}{r} - \binom{n - k}{r}$$

---

†See Glossary for a statement. For more details, see *Introduction to Inequalities* by E. Beckenbach and R. Bellman. NML 3, pp. 67–68.

‡See Glossary, *Series, Summation of-*.

to change (1) into

$$(3) \quad \sum_{k=1}^{n-r+1} \left[(k-1)\binom{n-k+1}{r} - k\binom{n-k}{r} \right] + \sum_{k=1}^{n-r+1} \binom{n-k+1}{r}.$$

By (2), the first sum in (3) is zero. To obtain the remaining sum, we again break up the summand as before, i.e.

$$\sum_{k=1}^{n-r+1} \binom{n-k+1}{r} = \sum_{k=1}^{n-r+1} \left[\binom{n-k+2}{r+1} - \binom{n-k+1}{r+1} \right]$$

$$= \binom{n+1}{r+1} = N(n,r).$$

Therefore

$$F(n,r) = \frac{\binom{n+1}{r+1}}{\binom{n}{r}} = \frac{(n+1)}{(r+1)}.$$

We now add a second solution which has the virtue that it avoids all calculations.

Second solution. A natural way to phrase the problem is this: Select r out of the first n positive integers at random. What is the expected value of the smallest integer selected?

The problem is discussed in books on probability and statistics under the heading "order statistics". One can formulate a simple solution without explicit reference to probabilities as follows.

Take a circle and divide it into $n+1$ arcs, numbered $0, 1, \ldots, n$. Take 1 red and r green markers. Consider the ways of placing the markers on the arcs so that no arc has more than 1 marker. For each possible position of the red marker each way of placing the green markers corresponds to a subset of r elements of $\{1, \ldots, n\}$, namely the distances of the green markers from the red one, measured, say, in the clockwise direction.

We want to find the *average distance* from the red marker to the next marker in the clockwise direction. We consider the configurations in groups of $r+1$, each group consisting of all configurations in which a given set of sectors is occupied. There are $r+1$ possible places for the red marker when the occupied sectors are given. The sum of the distances from the red marker to the next marker, taken over all these possibilities, is just the sum of the distances from one occupied sector to the next taken around the whole circle, which is $n+1$. Hence the average distance over each of these groups is $\dfrac{n+1}{r+1}$, which must also be the overall average.

There is an important principle governing the choice of the model used in the second solution above. We state it as a

THEOREM. *Let* $s = \{X\}$ *be the set of all configurations of some specified type* (*in our case, the selection of r integers from* $1, 2, \ldots, n$). *Let* $f(X)$ *be a function defined on the X's* (*in our case, f assigns to X its smallest member*). *We seek the average* $\frac{1}{k} \sum_{X \in S} f(X)$ *of f over S.*

Partition S into subsets S_j, so that the average value of f over each subset is $\bar{f}$:

$$(*) \qquad \frac{1}{k_j} \sum_{X \in S_j} f(X) = \bar{f},$$

where k_j is the number of elements in S_j,

$$\sum k_j = k.$$

Then the average of f over all $X \in S$ is $\bar{f}$.

PROOF: Simply multiply $(*)$ by k_j and then sum over j.

$$k_j \bar{f} = \sum_{X \in S_j} f(X), \quad \sum k_j \bar{f} = k\bar{f}, \quad \bar{f} = \frac{1}{k} \sum_{X \in S} f(X).$$

In our second solution, this principle is applied to the set consisting of all sets of $r + 1$ (out of $n + 1$) arcs, one marked red.

1981 / 3. We shall call an ordered pair (n, m) *admissible* if $m, n \in \{1, 2, \ldots, 1981\}$ and

$$(1) \qquad (n^2 - nm - m^2)^2 = 1.$$

If $m = 1$, then $(1, 1)$ and $(2, 1)$ are the only admissible pairs.

For any admissible pair (n_1, n_2) with $n_2 > 1$ we have

$$n_1(n_1 - n_2) = n_2^2 \pm 1 > 0,$$

so that $n_1 > n_2$. Define

$$n_3 = n_1 - n_2;$$

then $n_1 = n_2 + n_3$, and substituting this into (1), we have

$$1 = \left(n_1^2 - n_1 n_2 - n_2^2\right)^2$$
$$= \left((n_2 + n_3)^2 - (n_2 + n_3)n_2 - n_2^2\right)^2$$
$$= \left(-n_2^2 + n_2 n_3 + n_3^2\right)^2 = \left(n_2^2 - n_2 n_3 - n_3^2\right)^2$$

so (n_2, n_3) is also an admissible pair.

If $n_3 > 1$, then, in the same way, we conclude that $n_2 > n_3$; and, letting $n_2 - n_3 = n_4$ we find that (n_3, n_4) is an admissible pair.

Thus we have a sequence $n_1 > n_2 > n_3 > \dots$ (necessarily finite) such that

$$n_{i+1} = n_{i-1} - n_i,$$

and where (n_i, n_{i+1}) is admissible for all i.

The sequence terminates if $n_i = 1$. Since $(n_{i-1}, 1)$ is admissible, and $n_{i-1} > 1$, $n_{i-1} = 2$ must hold (see second sentence above). Therefore, (n_1, n_2) are consecutive terms of the truncated Fibonacci sequence 1597, 987, ..., 13, 8, 5, 3, 2, 1. Conversely, any such pair is admissible.

Every step of this construction is reversible; so, running it backwards from $(2, 1)$ determines uniquely the Fibonacci sequence $1, 2, 3, 5, 8, \dots$ which contains, as adjacent members, all admissible pairs. The largest such pair not exceeding 1981 is 1597, 987; so the maximum value of $m^2 + n^2$ is $1597^2 + 987^2$.

1981 / 4. The answers to (a) and (b) are $n \geqslant 4$ and $n = 4$, respectively. To see this, denote by N the largest number in a set of n consecutive positive integers. Note that in such a set there is exactly one member divisible by n.

Suppose $n = 3$. In order to divide the l.c.m.† of $N - 1$ and $N - 2$, N cannot have a prime factor $p > 2$ (since then p would not be a factor of either $N - 1$ or $N - 2$). So N is of the form 2^a where, since $N \geqslant 3$, a is an integer > 1. But N cannot be of this form because it must divide the odd number $N - 1$. This rules out $n = 3$.

Suppose $n = 4$. Then the l.c.m. of its three predecessors contains the factor $2^b 3^c$ for some positive integers b, c. Again, N could not have $p \geqslant 5$ as prime factor since, if it did, none of its predecessors could have the factor p, and so N could not divide their l.c.m. Therefore

$$N = 2^b 3^c, \quad \text{so} \quad N - 2 = 2(2^{b-1} 3^c - 1)$$

and

$$N - 3 = 3(2^b 3^{c-1} - 1).$$

Thus N divides their l.c.m. only if $b = 1$, $c = 1$, i.e. only if $N = 6$. Indeed, for the sequence $3, 4, 5, 6$, l.c.m. $(3, 4, 5) = 60$ is divisible by 6. This answers question (b).

Now suppose $n > 4$. We show that $(n - 1)(n - 2)$ and $(n - 2)(n - 3)$, both larger than n, are admissible values of N.

Any two consecutive integers are relatively prime. Thus, their l.c.m. is just their product. The n consecutive integers ending with N are

(1) $N - (n - 1), \quad N - (n - 2), \quad N - (n - 3), \quad \dots, \quad N - 1, \quad N.$

When $N = (n - 1)(n - 2)$, it divides the product of the first two integers in (1) because $n - 1$ divides $N - (n - 1) = (n - 1)(n - 3)$, and

---

†l.c.m. means least common multiple.

$n - 2$ divides $N - (n - 2) = (n - 2)(n - 2)$. Therefore, $(n - 1)(n - 2)$ divides the l.c.m. of the $n - 1$ integers preceding N.

When $N = (n - 2)(n - 3)$, we see easily that it divides the product of the second and third integers in (1); hence also $(n - 2)(n - 3)$ is an admissible N.

Thus for $n > 4$ we can find more than one sequence of n consecutive integers with a last member that divides the l.c.m. of all the others. This answers part (a) of the problem.

1981 / 5. Denote the vertices of the triangle by A, B, C and the centers of the circles by A', B', C'; see figure.

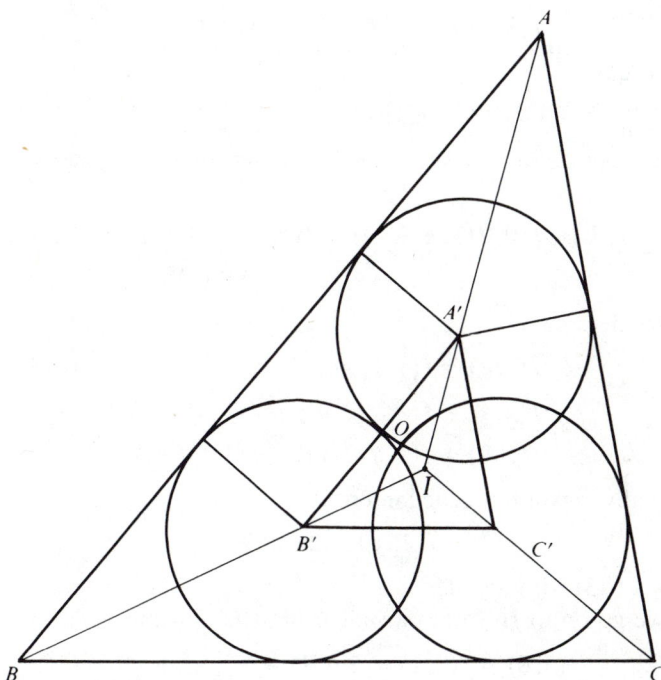

Since the three circles have equal radii, corresponding sides of triangles ABC and $A'B'C'$ are parallel, and O is the circumcenter, of $\triangle A'B'C'$. Hence $\triangle A'B'C'$ and $\triangle ABC$ are homothetic.† An important property of homothetic figures is that corresponding points lie on a line passing through the center of similitude (or homothety).‡ Lines AA', BB' and CC' bisect angles A, B, and C; they also connect corresponding points of the

†See Glossary.

‡For a proof, see e.g. *Geometry Revisited* by H. S. M. Coxeter and S. Greitzer, NML vol. 19, p. 94.

homothetic triangles, and they meet at the incenter I of $\triangle ABC$ (which is also the incenter of $\triangle A'B'C'$. So I is the center of similitude. The circumcenters of both triangles are corresponding points, so they lie on a line through I.

1981 / 6. We set $x = 0$ in conditions (2) and (1) to find
$$f(1,0) = f(0,1) = 2$$
and also set $y = 0$ in (3) and (1) to determine
$$f(1,1) = f(0, f(1,0)) = f(1,0) + 1 = 3.$$
Then, setting $y = 1$ in (3) and using the above gives
$$f(1,2) = f(0, f(1,1)) = f(0,3) = 4.$$
We claim that
$$(4) \qquad\qquad f(1, y) = y + 2.$$
We have verified this for $y = 0,1,2$ and see that it follows inductively from (3):
$$f(1, k) = f(0, f(1, k - 1)) = f(1, k - 1) + 1$$
$$= (k - 1) + 2 + 1 = k + 2.$$
Next we determine
$$f(2,0) = f(1,1) = 3 \qquad\qquad \text{by (3) and (4)},$$
$$f(2,1) = f(1, f(2,0)) = f(1,3) = 5,$$
$$f(2,2) = f(1, f(2,1)) = f(1,5) = 7$$
and generally, again by induction on y,
$$(5) \qquad\qquad f(2, y) = 2y + 3,$$
by means of (3) with $x = 1$.

Next we go on to $f(3, y)$ in a similar inductive manner:
$$f(3,0) = f(2,1) = 5 = 2^3 - 3$$
$$f(3,1) = f(2, f(3,0)) = 2 \cdot 5 + 3 = 2^4 - 3$$
and, in general,
$$(6) \qquad\qquad f(3, y) = 2^{y+3} - 3.$$
Finally, to find $f(4, 1981)$, we use
$$f(4, y + 1) = f(3, f(4, y))$$
and $f(4,0) = f(3,1) = 2^4 - 3,$
$$f(4,1) = f(3, f(4,0)) = 2^{2^4 - 3 + 3} = 2^{2^{2^2}},$$

and generally

$$f(4, y) = 2^{2^{\cdot^{\cdot^{\cdot^{2}}}}} - 3,$$

where $y + 3$ two's appear in the first term. Thus

$$f(4, 1981) = 2^{2^{\cdot^{\cdot^{\cdot^{2}}}}} - 3.$$

where 1984 two's appear in the first term.

Note. This function, defined by a double recursion, is called *Ackermann's function*. Ackermann, a student of Hilbert's, showed that it grows faster than any function that can be defined recursively in terms of a single variable. It has been of interest to computer scientists because it grows faster than anything programmable by simple loops. See [11, vol. 1, pp. 76–78].

Twenty-third International Olympiad, 1982

1982 / 1. The given relation implies that

$$f(m + n) \geqslant f(m) + f(n).$$

When $m = n = 1$, we have $f(2) \geqslant 2f(1)$; but $f(2) = 0$, and since the range of f is non-negative, we find that $f(1) = 0$.

When $m = 2$, $n = 1$, we have

$$f(3) = f(2) + f(1) + \{0 \text{ or } 1\} = 0 \text{ or } 1,$$

and since $f(3) > 0$, $f(3) = 1$. Then

$$f(2 \cdot 3) = f(3 + 3) \geqslant 2f(3) = 2;$$

so, for multiples of 3 we see by induction on n that

$$f(n \cdot 3) = f((n - 1) \cdot 3 + 3) \geqslant (n - 1)f(3) + f(3) = n,$$

so $f(3n) \geqslant n$ for all n. Moreover, if the strict inequality $>$ held for some n, i.e. if $f(3n) > n$, then also $f(3(n + 1)) > n + 1$, etc. so $f(3m) > m$ for all $m \geqslant n$. But since $f(9999) = f(3 \cdot 3333) = 3333$, the $=$ holds at least up to $n = 3333$: $f(3n) = n$ for $n \leqslant 3333$. In particular, $f(3 \cdot 1982) = 1982$.

On the other hand,

$$1982 = f(3 \cdot 1982) \geqslant f(2 \cdot 1982) + f(1982) \geqslant 3f(1982),$$

so

$$f(1982) \leqslant \frac{1982}{3} < 661.$$

Also

$$f(1982) \geqslant f(1980) + f(2) = f(3 \cdot 660) = 660$$

i.e., $660 \leqslant f(1982) < 661$. We conclude that $f(1982) = 660$.

Note that there is a function with the given properties, for example $f(n) = \left\lfloor \dfrac{n}{3} \right\rfloor$, where $\lfloor x \rfloor$ denotes the integer part of x.

1982 / 2. We used homothetic triangles† to solve problem 1981/5; here, we do so again. If we could show that triangles $M_1 M_2 M_3$ and $S_1 S_2 S_3$ are homothetic, then the desired result would be an immediate consequence. We know that triangles $M_1 M_2 M_3$ and $A_1 A_2 A_3$ are homothetic, and shall show that $S_1 S_2 S_3$ and $A_1 A_2 A_3$ are. Then the homothety of triangles $M_1 M_2 M_3$ and $S_1 S_2 S_3$ will follow, and their congruence can be ruled out.

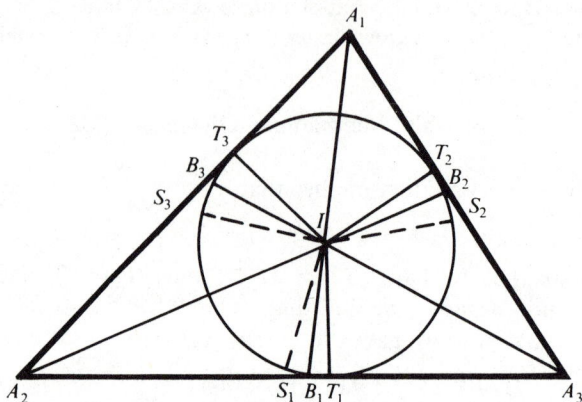

In the figure we have included the bisectors of $\angle s$ A_1, A_2, A_3 which intersect the opposite sides in points denoted by B_1, B_2, B_3, but we have omitted the midpoints M_i. We have

$$\angle T_1 I T_3 = 180° - A_2, \qquad \angle T_3 B_3 A_3 = A_2 + \tfrac{1}{2} A_3$$

$$\angle T_3 I B_3 = 90° - \left(A_2 + \tfrac{1}{2} A_3 \right),$$

$$\angle T_3 I S_3 = 2 \angle T_3 I B_3 = 180° - 2 A_2 - A_3.$$

Thus

$$\angle S_3 I T_1 = \angle T_3 I T_1 - \angle T_3 I S_3 = A_2 + A_3,$$

and similarly $\angle S_2 I T_1 = A_3 + A_2$. Hence $S_2 S_3 \| A_2 A_3$. Similarly $S_3 S_1 \| A_3 A_1$ and $S_1 S_2 \| A_1 A_2$. Triangles $S_1 S_2 S_3$ and $M_1 M_2 M_3$ therefore have corresponding sides parallel. They are not congruent since S_1, S_2, S_3 lie on the incircle of $\triangle A_1 A_2 A_3$ while M_1, M_2, M_3 lie on its nine point circle. These have different radii because $\triangle A_1 A_2 A_3$ is not equilateral. We conclude that triangles $S_1 S_2 S_3$ and $M_1 M_2 M_3$ are homothetic, so $M_1 S_1, M_2 S_2, M_3 S_3$ concur.

†See p. 35 and footnotes on that page.

1982 / 3. (a) We will prove that the series

(1) $\quad \dfrac{x_0^2}{x_1} + \dfrac{x_1^2}{x_2} + \dfrac{x_2^2}{x_3} + \cdots,\quad$ where $1 = x_0 \geqslant x_1 \geqslant x_2 \geqslant \cdots > 0$,

has sum $\geqslant 4$ (with the obvious convention that this holds if the series diverges). This clearly implies that some partial sum of the series is $\geqslant 3.999$.

Let L be the inf ($=$ greatest lower bound) of the sums of all series of the form (1). Clearly $L > 1$ since the first term $\dfrac{1}{x_1} \geqslant 1$. For any $\varepsilon > 0$, we can find a sequence $\{x_n\}$ such that

(2) $\qquad\qquad L + \varepsilon > \dfrac{x_0^2}{x_1} + \dfrac{x_1^2}{x_2} + \dfrac{x_2^2}{x_3} + \cdots$

Setting $y_n = x_{n+1}/x_1$ $(n \geqslant 0)$, we note that $1 = y_0 \geqslant y_1 \geqslant y_2 \geqslant \cdots > 0$. The series on the right side of (2) can be written in the form

$$\frac{1}{x_1} + x_1\left(\frac{y_0^2}{y_1} + \frac{y_1^2}{y_2} + \frac{y_2^2}{y_3} + \cdots \right).$$

By the definition of L, the series in parentheses has sum $\geqslant L$. Hence from (2) we have

$$L + \varepsilon > \frac{1}{x_1} + x_1 L.$$

Applying the A.M.-G.M. inequality to the right side, we get $L + \varepsilon > 2\sqrt{L}$. Since this holds for all $\varepsilon > 0$, it follows that $L \geqslant 2\sqrt{L}$. Hence $L^2 \geqslant 4L$, and since $L > 0$, this implies that $L \geqslant 4$.

(b) Let $x_n = \dfrac{1}{2^n}$. Then

$$\sum_{n=0}^{\infty} \frac{x_n^2}{x_{n+1}} = \sum_{n=0}^{\infty} \frac{1}{2^{n-1}} = 4,$$

so all partial sums of the series are < 4.

1982 / 4. (a) The given expression may be rewritten as

$$x^3 - 3xy^2 + y^3 = (y - x)^3 - 3x^2 y + 2x^3$$
$$= (y - x)^3 - 3(y - x)x^2 + (-x)^3.$$

Hence, if pair (x, y) is a solution, then so is pair $(y - x, -x)$. By a similar algebraic manipulation, we can show that $(-y, x - y)$ is then also a solution. In other words, if (x, y) is a solution, then the transformation

(1) T: $\begin{aligned} x' &= -x + y \\ y' &= -x \end{aligned}$ or $T\begin{pmatrix} x \\ y \end{pmatrix} = \begin{pmatrix} -1 & 1 \\ -1 & 0 \end{pmatrix}\begin{pmatrix} x \\ y \end{pmatrix}$

produces another solution. [The reader may verify that T^2 produces the third solution obtained above, and that $T^3 = I$, the identity.] Moreover, these three pairs are distinct; for, if two were identical, say $x = x'$ and $y = y'$, then $x = y = 0$ which is excluded by the condition $n > 0$.

 (b) To show that there are no integer solutions to

$$x^3 - 3xy^2 + y^3 = 2891,$$

we first consider the solutions $\mod 3$. Here the equation reduces to

(2) $x^3 + y^3 \equiv 2(\mod 3) \equiv -1(\mod 3).$

Thus we have three cases:

 (i) $x \equiv -1$, $y \equiv 0$, (ii) $x \equiv 0$, $y \equiv -1$, (iii) $x \equiv 1$, $y \equiv 1$, all $\mod 3$. No other candidates for (x, y) can satisfy (2). For (i), $x = 3m - 1$, $y = 3n$, and substituting back in our original equation, we get

$$(3m - 1)^3 - 3(3m - 1)(3n)^2 + (3n)^3 = 2891 = 9 \cdot 321 + 2$$

which obviously cannot hold $\mod 9$. Similarly for case (ii).

 The third case cannot occur, since it was shown that if (x, y) is a solution, so is $(y - x, -x)$; here $(1, 1)$ leads to $(0, -1)$, which is the same as (ii).

 We have covered all possible cases and conclude that (2) has no solution.

1982 / 5 First solution. Since $AC = EC$, we deduce from the given proportion that $CM = EN$, and $\triangle BMC \cong \triangle DNE$. See Figure (i). Thus

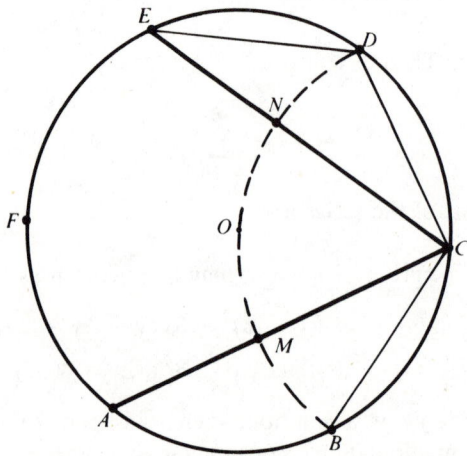

(i)

$\angle NBC = \angle EDN$. Since $\angle ECB = 90°$ and $\angle CED = 30°$, we have

$$\angle BND = \angle BNC + \angle CND$$
$$= (90° - \angle NBC) + (\angle CED + \angle NDE)$$
$$= 90° - \angle NBC + 30° + \angle NBC = 120°$$

Thus, BD subtends an angle of $120°$ from N and also from the center O of the circumcircle of the hexagon. So N lies on the circle with center C and radius $CD = CB = CN$. In right $\triangle BCE$, $\angle EBC = 60°$; so

$$r = \frac{CN}{CE} = \frac{CB}{CE} = \frac{1}{\sqrt{3}}.$$

Second solution. Assume that each side of the hexagon has length one, and that B, M, and N are collinear. Let X denote the intersection of AC and BE, see Figure (ii). Since N is on CE, B on EX and M on XC, we may apply Menelaus' Theorem,† to triangle CEX:

$$\frac{CN}{NE} \cdot \frac{EB}{BX} \cdot \frac{XM}{MC} = -1.$$

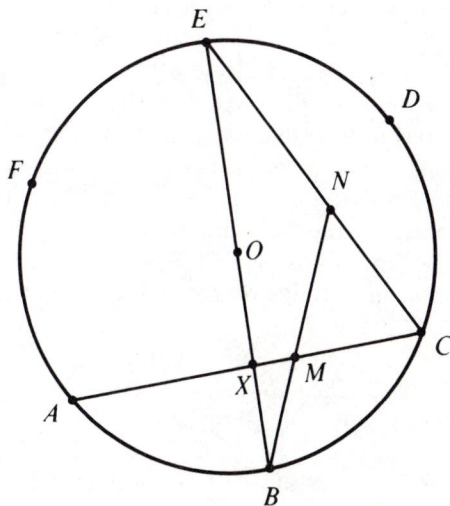

(ii)

We now obtain each of the distances in this formula in terms of r. CE is the side opposite the $120°$ angle in an isosceles triangle with two sides of

†See Glossary. For a proof, see e.g. *Geometry Revisited*, by H. S. M. Coxeter and S. L. Greitzer, NML vol. 19, p. 66.

lengths 1; so we have

$$CE = \sqrt{3}, \qquad Cn = \sqrt{3}\,r, \qquad EB = 2, \qquad BX = -\tfrac{1}{2}, \qquad AM = \sqrt{3}\,r$$

$$MC = AC - AM = \sqrt{3}\,(1 - r),$$

$$XM = \frac{\sqrt{3}}{2} - MC = \frac{\sqrt{3}}{2} - \sqrt{3} + r\sqrt{3} = \sqrt{3}\left(r - \tfrac{1}{2}\right).$$

Substituting into Menelaus' formula, we have

$$\frac{\sqrt{3}\,r}{\sqrt{3}\,(1 - r)} \cdot \frac{2}{-\tfrac{1}{2}} \cdot \frac{\sqrt{3}\left(r - \tfrac{1}{2}\right)}{\sqrt{3}\,(1 - r)} = -1,$$

and this yields $r = \dfrac{1}{\sqrt{3}}$.

1982 / 6. Denote the distance between a point P on the side of the square and a point Q on the polygonal path L by $d(PQ)$, and denote the length of the polygonal path from point A to point B by $s(AB)$. When

$$s(A_0 A) < s(A_0 B)$$

we shall write $A < B$. Let S_1, S_2, S_3, S_4 be the vertices of the square. On L, take points S_1', S_2', S_3', S_4' such that $d(S_i S_i') \leqslant \tfrac{1}{2}$. We may name the points so that $S_1' < S_4' < S_2'$; see Note below.

Now let L_1 be the set of all X on L such that $X \leqslant S_4'$, and let L_2 be the set of all X on L such that $X \geqslant S_4'$. Consider side $S_1 S_2$. There is a subset L_1' of $S_1 S_2$ whose points are distant $\leqslant \tfrac{1}{2}$ from L_1, and a subset L_2' of $S_1 S_2$ whose points are distant $\leqslant \tfrac{1}{2}$ from L_2. (Since L_1' includes S_1 and L_2' includes S_2, neither set is empty.)

The union of L_1' and L_2' is the side $S_1 S_2$, and, because of the condition on the distances, the intersection of L_1' and L_2' is not empty. Let M be a point common to L_1' and L_2'.

Now select a point X in L_1 and a point Y in L_2 such that $d(MX) \leqslant \tfrac{1}{2}$ and $d(MY) \leqslant \tfrac{1}{2}$. Then $d(XY) \leqslant 1$. Also $X < S_4' < Y$, and $s(XY) = x(XS_4') + s(S_4'Y) \geqslant 99 + 99 = 198.41$

Note. We can think of the path L: $A_0 A_1 \ldots A_n$, as a mapping Q of the interval $0 \leqslant t \leqslant 1$ into L, with $Q(0) = A_0$, $Q(1) = A_n$, and $Q(0) \neq Q(1)$. As we traverse L starting at $Q(0)$, there will be a first time, t_1, that L is within $\tfrac{1}{2}$ of a vertex of the square. Call that vertex S_1, and call $Q(t_1)$ the point S_1'. Denote by S_3 the vertex diagonally opposite S_1. Now continue traversing L until it gets within $\tfrac{1}{2}$ of one of the vertices adjacent to S_1. Call that adjacent vertex S_4, and denote a point on L within $\tfrac{1}{2}$ of S_4 by $S_4' = Q(t_4)$. S_1 and S_4 are endpoints of a side of the square, they orient the boundary in the order $S_1 S_4 S_3 S_2$, and L must approach S_2 eventually, for values of the parameter $t > t_4$. Select a point on L within $\tfrac{1}{2}$ of S_2, call it $S_2' = Q(t_2)$, and the desired naming is accomplished.

Twenty-fourth International Olympiad, 1983

1983 / 1. We first show that 1 is in the range of f. For an arbitrary $x_0 > 0$, let

$$y_0 = \frac{1}{f(x_0)}.$$

Then the given equation (i) yields

$$f(x_0 f(y_0)) = y_0 f(x_0) = 1,$$

so 1 is in the range of f. [The same argument shows that any positive real number is in the range of f.] Hence there is a value y such that $f(y) = 1$. This together with $x = 1$ in (i) gives

$$f(1 \cdot 1) = f(1) = yf(1).$$

Since $f(1) > 0$ by hypothesis, it follows that $y = 1$, and

$$f(1) = 1.$$

When we set $y = x$ in (i), we get

$$(1) \qquad\qquad f(xf(x)) = xf(x) \qquad\qquad \text{for all } x > 0.$$

Hence $xf(x)$ is a fixed point of f. Now if a and b are fixed points of f, that is, if

$$f(a) = a, \qquad f(b) = b,$$

then (i) with $x = a$, $y = b$ implies that

$$f(ab) = ba,$$

so ab is also a fixed point of f. Thus the set of fixed points of f is closed under multiplication. In particular, if a is a fixed point, all non-negative integral powers a^n of a are fixed points. Then since by (ii) $f(x) \to 0$ as $x \to \infty$, there can be no fixed points > 1. Since $xf(x)$ is a fixed point, it follows that

$$(2) \qquad\qquad xf(x) \leqslant 1, \qquad \text{or} \qquad f(x) \leqslant \frac{1}{x} \qquad\qquad \text{for all } x.$$

Let

$$a = zf(z), \qquad \text{so} \qquad f(a) = a.$$

Now set $x = 1/a$ and $y = a$ in (i) to give

$$f\left(\frac{1}{a}f(a)\right) = f(1) = 1 = af\left(\frac{1}{a}\right)$$

or

$$f\left(\frac{1}{a}\right) = \frac{1}{a}, \quad \text{or} \quad f\left(\frac{1}{zf(z)}\right) = \frac{1}{zf(z)}.$$

This shows that $1/xf(x)$ is also a fixed point of f for all $x > 0$. Therefore

$$f(x) \geq \frac{1}{x}.$$

This together with (2) implies that

(3) $$f(x) = \frac{1}{x}.$$

The function (3) is the only solution satisfying the hypothesis.

1983 / 2. Let O be the intersection of P_2P_1, Q_2Q_1 and O_2O_1. The two circles are homothetic and O is their center of similitude.† Let B be the other point of intersection of the circles, T be the intersection of AB with P_2P_1, and C be the other point of intersection of OA with C_2. Since $TA \cdot TB = TP_2^2 = TP_1^2$, TA is the perpendicular bisector of M_2M_1. Hence $\alpha = \angle AM_1O_1 = \angle AM_2O_1 = \angle BM_2O_1$. By homothety $\alpha = \angle CM_2O_2$. Hence M_2, C and B are collinear. Now by reflection $\beta = \angle O_2AM_2 = \angle O_2BM_2$. From the isosceles triangle CO_2B we get $\angle O_2CM_2 = \beta$, and by homothety $\angle O_1AM_1 = \beta$. Thus both angles in the original question are equal to $\beta + \angle M_2AO_1$.

†See Glossary.

1983 / 3. Assume that $2abc - bc - ca - ab$ is expressible in the form $xbc + yca + zab$. Since $(a, b) = (a, c) = 1$, a must divide $x + 1$. Thus $x + 1 \geqslant a$, and similarly $y + 1 \geqslant b$, $z + 1 \geqslant c$. However, this leads to the contradiction

$$2abc = (x + 1)bc + (y + 1)ca + (z + 1)ab \geqslant abc + bca + cab = 3abc.$$

To show that every $n > 2abc - bc - ca - ab$ is expressible in the form

$$xbc + yca + zab, \qquad\qquad x, y, z \geqslant 0,$$

we first solve a simpler problem, involving only two natural, relatively prime numbers, b and c: Find the largest integer n which *cannot* be represented as

$$n = u'b + v'c \qquad\qquad \text{with integers } u' \geqslant 0, \ v' \geqslant 0.$$

To solve this problem, we determine *all* numbers m which cannot be so represented. Now since $(b, c) = 1$, every integer m can be represented as

$$m = ub + vc, \qquad\qquad u, v \text{ integers.†}$$

This representation is not unique; but all others are of the form

$$m = (u - kc)b + (v + kb)c, \qquad k \text{ any integer.}$$

Is there, among these representations, one where

$$u - kc \geqslant 0, \qquad v + kb \geqslant 0 \ ?$$

To decide this, choose k as large as possible, so that $u - kc \geqslant 0$; if, for this choice of k, $v + kb \geqslant 0$, then m can be represented as linear combination of b and c with non-negative coefficients. Clearly, for the largest choice of k,

$$0 \leqslant u - kc < c.$$

Set $u' = u - kc$, $v' = v + kb$, and write

$$(1) \qquad\qquad m = u'b + v'c;$$

call this a *reduced* representation if $0 \leqslant u' < c$.

We have shown

(a) Every integer m has a reduced representation.

(b) The integer m can be represented with non-negative coefficients if and only if its reduced representation (1) has $v' \geqslant 0$.

This tells us that the numbers m which cannot be represented as desired are of the form (1) with

$$0 \leqslant u' < c, \qquad v' < 0;$$

---

†For a proof, see e.g. C. D. Olds, *Continued Fractions*, NML vol. 9 p. 36 ff.

the largest among these has

$$u' = c - 1, \qquad v' = -1,$$

so the largest is

(2) $$m = (c - 1)b - c = bc - b - c.$$

We now return to the given problem and shall make use of the above result twice. Since $(b, c) = 1$, we have $b > 1$, $c > 1$, and

$$b(c - 1) > c - 1, \quad \text{or} \quad bc - b - c > -1.$$

Multiply both members by $a > 0$, and add $abc - bc$ to both sides, obtaining

$$2abc - ab - ac - bc > abc - a - bc.$$

Now if n is an integer such that

(3) $$n > 2abc - bc - ca - ab,$$

then also

(4) $$n > a(bc) - a - (bc).$$

This inequality and the fact that $(a, bc) = 1$ allow us to apply our previous result (with bc now playing the role of b in (1) and a that of c); i.e. n is large enough to be expressed as

(5) $$n = xbc + pa \quad \text{with } 0 \leqslant x \leqslant a - 1, \quad p > 0.$$

From (5), we have

$$pa = n - xbc \geqslant n - (a - 1)bc = n - abc + bc,$$

and from (3),

$$n - abc + bc > abc - ab - ac.$$

Thus

$$pa > abc - ab - ac \quad \text{or} \quad p > bc - b - c.$$

Another application of our solution to the simpler problem gives the representation $p = yc + zb$ with integers y, $z \geqslant 0$. Setting this into (5), we find

$$n = (yc + zb)a + xbc = xbc + yca + zab, \qquad x, y, z \geqslant 0.$$

This problem is given as an exercise in H. L. Keng, *Introduction to Number Theory*, Springer-Verlag, Heidelberg, (1982) pp. 11–12, together with the following unsolved variant: If a, b, c are positive integers with $(a, b, c) = 1$, determine the largest integer not representable as $ax + by + cz$ with $x, y, z > 0$.

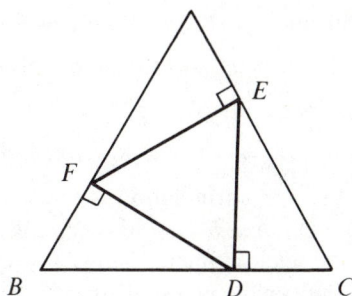

1983 / 4. We will show that the answer is in the affirmative. Our proof is indirect. First we choose points D, E, F on BC, CA, AB respectively, such that $BD/DC = CE/EA = AF/FB = 2$. It follows easily that AFE, BDF, and CED are 60°-30°-90° triangles. We now assume there exists a two-coloring of the points of $\mathcal{E}$, say red and blue, so that there are no monochromatic right triangles. Two of the points D, E, F must have the same color; say E and F are red. Then every point in $AC - [E]$ must be blue. This in turn implies that every point P in $AB \cup BC - [A, C]$ must be red, since otherwise the triangle APQ, where Q is the orthogonal projection of P onto AC, would be monochromatic. But then triangle BFD is monochromatic. This gives a contradiction, and so every two-coloring of $\mathcal{E}$ must contain a monochromatic right triangle.

1983 / 5. We construct a set T containing even more than 1983 integers, all less than 10^5 such that no three are in arithmetic progression, that is, no three satisfy $x + z = 2y$.

The set T consists of all positive integers whose base 3 representations have at most 11 digits, each of which is either 0 or 1 (i.e., no 2's). There are $2^{11} - 1 > 1983$ of them, and the largest is

$$1 + 3^2 + 3^3 + \cdots + 3^{10} = 88573 < 10^5.$$

Now suppose $x + z = 2y$ for some $x, y, z \in T$. The number $2y$, for any $y \in T$, consists only of the digits 0 and 2. Hence x and z must match digit for digit, and it follows that $x = z = y$. Hence T contains no arithmetic progressions of length 3, and the desired selection is possible.

1983 / 6 First solution. In establishing triangle inequalities, it is often helpful to use the transformation $a = y + z$, $b = z + x$, $c = x + y$, where x, y, z are arbitrary non-negative numbers. Inversely

$$x = \tfrac{1}{2}(b + c - a) = s - a, \quad y = s - b, \quad z = s - c,$$

$$\text{where } s = \tfrac{1}{2}(a + b + c).$$

Then for any triangle inequality $I(a, b, c) \geqslant 0$, we have

$$I(a, b, c) \geqslant 0 \Leftrightarrow I(y + z, z + x, x + y) \geqslant 0.$$

Dually, for any x, y, z inequality $J(x, y, z) \geqslant 0$, we have

$$J(x, y, z) \geqslant 0 \Leftrightarrow J(s - a, s - b, s - c) \geqslant 0.$$

One advantage of the x, y, z formulation is that the bothersome triangle constraints $b + c > a$, $c + a > b$, $a + b > c$, are transformed into the simple statement that the semi-perimeter minus any side is positive.† In the x, y, z formulation, the given inequality reduces to

(1) $$xy^3 + yz^3 + zx^3 \geqslant xyz(x + y + z).$$

To prove (1), we use Cauchy's inequality‡ in the following manner:

$$\left(xy^3 + yz^3 + zx^3\right)(z + x + y) \geqslant \left(y\sqrt{xyz} + z\sqrt{xyz} + x\sqrt{xyz}\right)^2$$
$$= xyz(x + y + z)^2.$$

There is equality if and only if $(xy^3, yz^3, zx^3) = k(z, x, y)$. Thus $x = y = z$, i.e. the triangle is equilateral.

Second solution. (due to Bernhard Leeb) After some algebraic manipulations, the given inequality can be rewritten in the form

(2) $$a(b - c)^2(b + c - a) + b(a - b)(a - c)(a + b - c) \geqslant 0.$$

Since a cyclic permutation of (a, b, c) leaves the given inequality unchanged, we can assume without loss of generality that $a \geqslant b, c$; (2) is valid since $b + c - a$ (or x) and $a + b - c$ (or z) are > 0. There is equality if and only if $a = b = c$.

In both solutions, the equality holds in (1) and (2) also when the triangle is degenerate, a circumstance ruled out in the problem statement.

Twenty-fifth International Olympiad, 1984

1984 / 1 First solution. Since x, y, z are non-negative and $x + y + z = 1$, at least one of these three numbers, say z, is $\leqslant \frac{1}{2}$. Thus the given expression

$$G = yz + zx + xy - 2xyz = z(x + y) + xy(1 - 2z)$$

written as a sum of non-negative terms is non-negative. For the least value of G, one of x, y must be 1 and the other 0. Thus, $G \geqslant 0$.

---

†For further applications of the $a, b, c, -x, y, z$ duality, see Crux Mathematicorum 10 (1984) 46–48, 11 (1985) 179–180.

‡See Glossary.

By the A.M.-G.M. inequality,

$$x + y \geqslant 2\sqrt{xy}, \quad \text{so } (1 - z)^2 = (x + y)^2 \geqslant 4xy.$$

Therefore,

$$G - \frac{7}{27} = z(x + y) + xy(1 - 2z) - \frac{7}{27}$$

$$\leqslant z(1 - z) + \frac{1}{4}(1 - z)^2(1 - 2z) - \frac{7}{27}$$

$$= -\frac{(3z - 1)^2(6z + 1)}{108} \leqslant 0.$$

Thus, $0 \leqslant G \leqslant 7/27$. The maximum value of G is taken on only for $x = y = z = \frac{1}{3}$.

Second solution. In terms of homogeneous polynomials, the given left hand inequality is equivalent to

$$(x + y + z)(yz + zx + xy) \geqslant 2xyz, \quad \text{where } x, y, z \geqslant 0,$$

since $x + y + z = 1$. This inequality follows immediately from the known, sharper inequality

(1) $$(x + y + z)\left(\frac{1}{x} + \frac{1}{y} + \frac{1}{z}\right) \geqslant 9$$

which can be derived, for example, by applying the A.M.-G.M. inequality to each factor on the left, and which is equivalent to

(2) $$(x + y + z)(yz + zx + xy) \geqslant 9xyz.$$

In terms of the elementary symmetric functions

$$T_1 \equiv x + y + z, \qquad T_2 \equiv yz + zx + xy, \quad \text{and} \quad T_3 \equiv xyz$$

of x, y, z, (2) takes the form

$$T_1 T_2 \geqslant 9T_3.$$

The given right hand side inequality, $G \leqslant 7/27$, is equivalent to

$$(x + y + z)(yz + zx + xy) - 2xyz \leqslant \frac{7}{27}(x + y + z)^3$$

or to

$$7T_1^3 \geqslant 27T_1 T_2 - 54T_3.$$

It is known† that the "best" linear inequality in T_1^3, $T_1 T_2$, and T_3 is

(3) $$T_1^3 \geqslant 4T_1 T_2 - 9T_3.$$

†See R. Frucht, M. S. Klamkin, "On best quadratic triangle inequalities," *Geometriae Dedicata 2* (1973), 341–348.

By "best," we mean that if there is any other valid inequality of the form

$$T_1^3 \geqslant aT_1T_2 - bT_3,$$

then

$$4T_1T_2 - 9T_3 \geqslant aT_1T_2 - bT_3.$$

In particular

$$4T_1T_2 - 9T_3 \geqslant \frac{27}{7}T_1T_2 - \frac{54}{7}T_3,$$

and this reduces to $T_1T_2 \geqslant 9T_3$ and therefore to inequality (1).

For completeness, we establish (3). After some algebraic manipulation, (3) turns out to be the special case, $n = 1$, of Schur's inequality

(4) $x^n(x - y)(x - z) + y^n(y - z)(y - x) + z^n(z - x)(z - y) \geqslant 0$

for $n \geqslant 0$ and for all real x, y, z. We can assume without loss of generality that $x \geqslant y \geqslant z$. Then the inequality (4) follows from the two obvious inequalities

$$x^n(x - y)(x - z) \geqslant y^n(x - y)(y - z) \quad \text{and} \quad z^n(z - x)(z - y) \geqslant 0.$$

Remark. Many contestants used multivariate calculus in this problem to the chagrin of many jury members. However, it should be pointed out that calculus problems are not in the *unwritten* syllabus for IMO problems. Although an occasional IMO problem can be solved by means of calculus, so far all such problems could be done more simply by more elementary methods. Nevertheless, since many contestants are knowledgeable in the calculus and the method of optimization is a standard one, there is no reason why students should avoid it, especially if they do not see a more elementary solution at the time. But to get full credit, the contestant must establish all necessary and sufficient conditions.

1984 / 2 Solution (due to Jeremy Kahn). We have

$$(a + b)^7 - a^7 - b^7 = 7ab(a + b)(a^2 + ab + b^2)^2.$$

Since 7 does not divide $ab(a + b)$, we must choose a, b so that 7^3 divides $a^2 + ab + b^2$, i.e. so that

(1) $$a^2 + ab + b^2 \equiv 0 (\bmod 7^3).$$

Since

(2) $$a^3 - b^3 = (a - b)(a^2 + ab + b^2),$$

(2) is equivalent to

(3) $$a^3 \equiv b^3 (\bmod 7^3).$$

By Euler's extension of Fermat's theorem,† for any number c relatively prime to n, we have

$$c^{\phi(n)} \equiv 1(\bmod\ n); \quad \text{now } \phi(7^3) = (7-1)7^2 = 3 \cdot 98,$$

so $c^{3 \cdot 98} \equiv 1(\bmod\ 7^3)$ for any $c \not\equiv 0(\bmod\ 7)$, e.g., for $c = 2$. Set $b = 1$, $a = 2^{98}$; then

$$(2^{98})^3 \equiv 1(\bmod\ 7^3).$$

Moreover, since $2^{98} \equiv 4(\bmod\ 7)$, $a + b = 2^{98} + 1 \equiv 5(\bmod\ 7)$ and $a - b = 2^{98} - 1 \equiv 3(\bmod\ 7)$, 7 does not divide $ab(a + b)$ nor $(a - b)$.

Now 2^{98} is terribly large and can certainly be reduced to manageable size (mod $7^3 = 343$). For example,

$$2^{10} = 1024 = 3 \cdot 7^3 - 5, \quad \text{so} \quad 2^{10} \equiv -5(\bmod\ 7^3), \quad 2^{20} \equiv 25(\bmod\ 7^3),$$

$$2^{40} \equiv 61, \quad 2^{80} \equiv -52, \quad 2^{90} \equiv -83, \quad 2^8 \equiv -87, \quad 2^{98} \equiv 18(\bmod\ 7^3).$$

So $a = 18$, $b = 1$ is a solution.

Remark. Many contestants solved this problem by educated trial and error as follows: First they established as above that (1) is satisfied. Also from $(a + b)^2 > a^2 + ab + b^2 \geqslant 7^3 = 343$, they concluded that $a + b > 18$, and then were very fortunate in that the first case, $a + b = 19$, worked with a or $b = 18$. This would have been a much better problem if it had asked for a family of solutions.

1984 / 3. Suppose every point of the plane has been colored with one of n different colors; then on any subset of points, $k \leqslant n$ colors may appear. There are altogether

$$\sum_{k=1}^{n} \binom{n}{k} = 2^n - 1$$

different color combinations that may appear on a pointset.

Now consider the set of all concentric circles with center O whose radii are less than 1. Among any 2^n members of this family of circles, there are at least two, call them R and S, bearing the same set of colors. We name them so that their radii r and s satisfy $0 < r < s < 1$.

We claim that there is a point Y on R such that the circle $C(Y)$, whose radius is

$$r + \frac{a(Y)}{r},$$

coincides with S; that is, such that

$$r + \frac{a(Y)}{r} = s \quad \text{or} \quad a(Y) = r(s - r).$$

†See Glossary.

Clearly, $0 < r(s - r) < 1$, and as point X moves along R in the counter-clockwise direction (starting on line OA), the measure of angle AOX covers every point in the interval $(0, 1)$. When this angle has measure $r(s - r)$, X has the desired position Y on R, so $C(Y) = S$. Moreover, the color of Y appears somewhere on circle S.

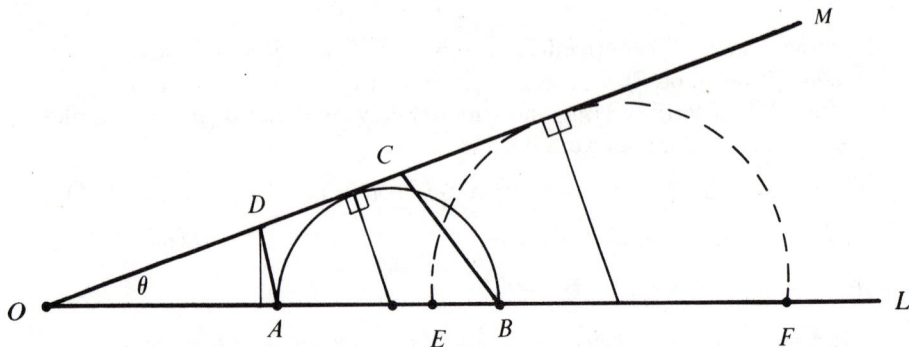

1984 / 4. Solution   (due to Loren Larson and Bruce Hanson)†. First we consider rays OL and OM making an acute angle $\theta = \angle LOM$ and two points E and F on one of the rays, say OL, with E between O and F; see figure. Let $C(E, F)$ denote the circle with diameter EF. Then, by similarity, $C(E, F)$ is tangent to OM if and only if the ratio

$$\frac{EF}{OF} = K(\theta)$$

is independent of the particular choice of E, F, i.e., if and only if $K(\theta)$ is a constant for fixed θ.

Now consider our problem and assume first that AB and CD are not parallel. Take OL and OM to be rays passing through A, B and C, D respectively. Since $C(A, B)$ is tangent to OM,

$$\frac{AB}{OB} = K(\theta);$$

and the circle $C(C, D)$ is tangent to OL if and only if

$$\frac{DC}{OC} = K(\theta).$$

These ratios, AB/OB and DC/OC, are equal if and only if BC is parallel to AD.

†This solution was published in Math Magazine, 58 (1985) p. 59.

In the case that AB and CD are parallel, the circle $C(C, D)$ is tangent to AB if and only if $AB = CD$, and this holds if and only if $ABCD$ is a parallelogram, i.e., if and only if BC and AD are also parallel.

1984 / 5. Consider the convex polygon $A_0 A_1 A_2 \ldots A_{n-1}$, with the subscripts reduced (mod n). Let $A_i A_j$ be a diagonal. Then by the triangle inequality,

$$A_i A_j + A_{i+1} A_{j+1} > A_i A_{i+1} + A_j A_{j+1}.$$

When we sum these inequalities over all $\frac{1}{2} n(n - 3)$ diagonals $A_i A_j$, each diagonal occurs twice on the left, and each side $n - 3$ times on the right; so we get

$$2d > (n - 3)p \qquad \text{or} \qquad n - 3 < \frac{2d}{p}.$$

To obtain the upper bound, consider the diagonal $A_i A_j$. Since it is the shortest path connecting A_i and A_j, it is shorter than each polygonal path joining its endpoints. Thus we have both:

$$A_i A_j < A_i A_{i+1} + \cdots + A_{j-1} A_j \quad \text{and} \quad A_i A_j < A_j A_{j+1} + \cdots + A_{i-1} A_i.$$

If n is odd, say $n = 2k - 1$, use for any diagonal $A_i A_j$ that of the two inequalities above with fewer terms on the right. When we sum these $\frac{1}{2} n(n - 3)$ inequalities, we get a d on the left, and on the right a sum of side lengths where each side appears exactly $2 + 3 + \cdots + k - 1 = \frac{1}{2} k(k - 1) - 1$ times. Therefore

$$d < \frac{p}{2} (k(k - 1) - 2) = \frac{p}{2} \left(\frac{n + 1}{2} \cdot \frac{n - 1}{2} - 2 \right).$$

If n is even, say $n = 2k$, use the same inequalities as above except for diagonals $A_i A_{i+k}$. For these k "diameters", use the inequalities $A_i A_{i+k} \leqslant \frac{1}{2} p$. Summing these $\frac{1}{2} n(n - 3)$ inequalities, we get

$$d < k \cdot \tfrac{1}{2} p + \tfrac{1}{2} p (k(k - 1) - 2) = \frac{p}{2} (k(k - 1) - 2 + k)$$

$$= \frac{p}{2} (k^2 - 2) = \frac{p}{2} \left(\frac{n^2}{4} - 2 \right).$$

Finally it is very easy to show that for even n, $\dfrac{n^2}{4} = \left[\dfrac{n}{2}\right]\left[\dfrac{n + 1}{2}\right]$, and for odd n, $\dfrac{n^2 - 1}{4} = \left[\dfrac{n}{2}\right]\left[\dfrac{n + 1}{2}\right]$, where $[x]$ denotes the integer part of x. Thus the upper bound for d in all cases can be written

$$d < \frac{p}{2} \left(\left[\frac{n}{2}\right]\left[\frac{n + 1}{2}\right] - 2 \right).$$

1984 / 6. Since $ad = bc$, we have

$$a((a + d) - (b + c)) = (a - b)(a - c) > 0,$$

so that $2^k = a + d > b + c = 2^m$, hence $k > m$. Since

$$ad = a(2^k - a) = bc = b(2^m - b),$$

we have

$$b^2 - a^2 = 2^m(b - 2^{k-m}a) = (b - a)(b + a);$$

therefore 2^m divides $(b - a)(b + a)$. The numbers $b - a$ and $b + a$ are not both divisible by 4, because their sum, $2b$, is not. We conclude that one of them is divisible by 2^{m-1}; denote it by x. Then

$$0 < x \leqslant b + a < b + c = 2^m,$$

so that we must have

$$x = 2^{m-1}.$$

The number $b + c - x = 2^{m-1}$ is equal to either $c + a$ (if $x = b - a$) or to $c - a$ (if $x = b + a$). If a, b had a common divisor, d, then d would divide also their sum (and their difference); but x is divisible only by 2, a, b are odd, so $(a, b) = 1$. Similarly, $(a, c) = 1$. However, a divides bc. Consequently $a = 1$.

Remark. With a little further work we can determine all a, b, c, d, satisfying the given conditions. Since

$$b - a < b < \tfrac{1}{2}(b + c) = 2^{m-1},$$

we have $b - a \neq 2^{m-1}$; hence the number x defined above is $b + a = 2^{m-1}$. Since $a = 1$, this gives $b = 2^{m-1} - 1$, $c = 2^m - b = 2^{m-1} + 1$, and finally

$$d = bc = 2^{2m-2} - 1.$$

Conversely, for any integer $m \geqslant 3$, the numbers $a = 1$, $b = 2^{m-1} - 1$, $c = 2^{m-1} + 1$, $d = (2^{m-1})^2 - 1$ satisfy the given conditions (with $k = 2(m - 1)$).

Twenty-sixth International Olympiad, 1985

1985 / 1. In the figure, O on side AB is the center of the circle, and E, F, G are the points of tangency. We rotate $\triangle OFC$ about O to get $\triangle OEH$, where H is on line AD. Set $\theta = \angle OCF = \angle OHE$; then also $\angle OCG = \theta$.

Since quadrilateral $ABCD$ is cyclic, $\angle OAH = \pi - 2\theta$, so that $\angle AOH = \pi - (\theta + \pi - 2\theta) = \theta = \angle AHO$. Hence

(1) $$OA = AH = AE + FC = AE + GC.$$

By the same argument, i.e., by rotating $\triangle OFD$ about O into $\triangle OGK$, K on line BC, etc., we find

(2) $$OB = BK = BG + GK = BG + ED.$$

Adding (1) and (2), we have $AB = AD + BC$.

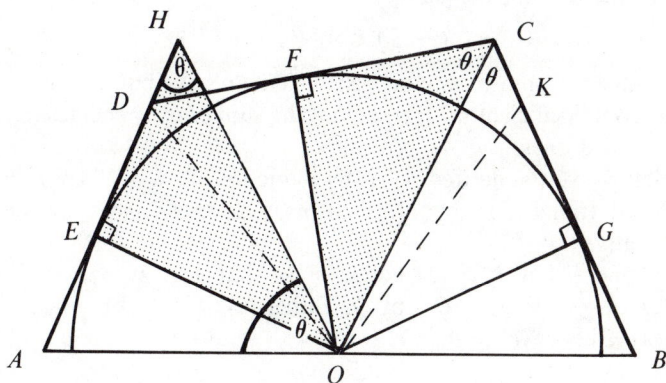

1985 / 2 Solution (due to Waldemar Horwat). To show that all elements have the same color, we shall permute them in such a way that the given conditions imply that each element in the permuted set has the same color as its predecessor.

We consider the first $n - 1$ multiples of k and reduce them (mod n), setting

$$m_r \equiv rk(\text{mod } n), \qquad r = 1, 2, \ldots, n - 1.$$

Since $(n, k) = 1$, $rk \not\equiv 0$, and since $ik \equiv jk(\text{mod } n)$ if and only if $i = j$, the set $\{m_1, m_2, \ldots, m_{n-1}\}$ is a permutation of M.

Now

(a) $$m_r = m_{r-1} + k \qquad \text{if } m_{r-1} + k < n,$$

and

(b) $$m_r = m_{r-1} + k - n \qquad \text{if } m_{r-1} + k > n.$$

In case (a), m_r has the same color as $|m_r - k| = m_{r-1}$ by (ii); in case (b), m_r has the same color as $|m_r - k| = n - m_{r-1}$ by (ii), which in turn has the same color as m_{r-1} by (i).

1985 / 3. We first observe that in the special case $k = 2^m$, all coefficients in the binomial expansion of $(1 + x)^k$ except the first and last are even. So

(1) $$(1 + x)^k \equiv 1 + x^k(\text{mod } 2).$$

Secondly, if R and S denote arbitrary polynomials with integer coefficients, and

(2)   if $\deg(R) < k$, then $w(R + x^k S) = w(R) + w(S)$,

since all terms in R are of lower powers of x than all the terms of $x^k S$. Moreover, the triangle inequality

$$w(R + S) \leqslant w(R) + w(S)$$

is valid, since the sum of any odd coefficients of like powers in R and S yields an even coefficient in $R + S$, but the sum of even coefficients cannot create new odd ones.

We shall prove the desired result by induction on i_n. When $i_n = 0$, the results holds trivially. If $i_n = 1$, then n is 2 or 1, and again the result holds trivially.

Now suppose that $i_n > 1$. Choose m so that $k = 2^m \leqslant i_n < 2^{m+1}$ and denote $Q_{i_1} + Q_{i_2} + \cdots + Q_{i_n}$ by Q. Either (a) $i_1 < k$, or (b) $i_1 \geqslant k$.

Case (a). Choose r so that $i_r < k \leqslant i_{r+1}$ and set

$$Q = Q_{i_1} + Q_{i_2} + \cdots + Q_{i_r} + Q_{i_{r+1}} + \cdots + Q_{i_n} = R + (1 + x)^k S,$$

where R is the sum of the first r Q_i's, and $\deg(R)$ and $\deg(S)$ are $< k$. Now

$$w(Q) = w(R + (1 + x^k)S)$$
$$= w(R + S + x^k S) \qquad\qquad \text{by (1)}$$
$$= w(R + S) + w(S) \qquad\qquad \text{by (2).}$$

Since $R = R + S - S$ and $w(-S) = w(S)$, the triangle inequality yields $w(R) \leqslant w(R + S) + w(S)$, so

$$w(Q) \geqslant w(R),$$

and by the induction hypothesis, $w(R) \geqslant w(Q_{i_1})$. Therefore, in case (a)

$$w\left(\sum_{j=1}^{n} Q_{i_j} \right) \geqslant w(Q_{i_1}).$$

Case (b). $i_1 \geqslant k = 2^m$, so we can write

$$Q_{i_1} = (1 + x)^k R \qquad \text{and} \qquad Q = (1 + x)^k S.$$

Since $i_n < 2^{m+1}$, $\deg(R)$ and $\deg(S)$ are $< k$. We again use (1) and (2) to get

$$w(Q) = w(S + x^k S) = 2w(S).$$

By the induction hypothesis, $w(S) \geqslant w(R)$, so

$$w(Q) \geqslant 2w(R) = w(R + x^k R) = w(Q_{i_1}).$$

This completes the solution.

1985 / 4. Since there are only 9 primes < 26, each of the 1985 members of M has a prime factorization in which at most 9 distinct primes occur:

$$(1) \quad m = p_1^{k_1} p_2^{k_2} \ldots p_9^{k_9}, \qquad k_i \text{ integers} \geqslant 0, \qquad \text{for each } m \in M.$$

To each member of M we assign the 9-tuple (i.e. the vector)

$$(x_1, x_2, \ldots, x_9)$$

where

  $x_i = 0$ if the exponent k_i of p_i in (1) is even,
  $x_i = 1$ if k_i is odd.

Thus 2^9 distinct vectors are possible. By the pigeonhole principle, any subset of $2^9 + 1$ elements of M contains at least two distinct integers, say a_1 and b_1, with the same exponent vector. It follows that their product is a perfect square: $a_1 b_1 = c_1^2$.

When we remove such a pair from the set M, we are left with $1985 - 2 > 2^9 + 1$ numbers, apply the pigeonhole principle again and continue removing such pairs as long as at least $2^9 + 1$ numbers are left in M. Since

$$1985 > 3(2^9 + 1) = 1539,$$

we can remove $2^9 + 1$ pairs a_i, b_i and still have $1985 - 2(2^9 + 1) = 959 > 2^9 + 1 = 513$ numbers left in M.

We now look at the $2^9 + 1$ removed pairs and take the square roots c_i of their products $a_i b_i = c_i^2$:

$$c_i = \sqrt{a_i b_i}.$$

The c_i cannot contain prime factors other than $p_1, p_2, \ldots, p_q$, so there is at least one pair, c_i, c_j with the same exponent vector, and $c_i c_j = d^2$ for some integer d. It follows that

$$d^4 = c_i^2 c_j^2 = a_i b_i a_j b_j$$

for some a_i, b_i, a_j, b_j in M.

Note that the same argument yields the more general result: *If the prime divisors of M are limited to n distinct primes, and if M contains at least $3(2^n + 1)$ distinct elements, then M contains a subset of four elements whose product is the fourth power of some integer.*

1985 / 5. The three circles, taken in pairs, have as radical axes the three lines through AC, through KN, and through BM. The center of the circumcircle of $\triangle ABC$ and O lie on the perpendicular bisector of AC, but the center of the circumcircle of $\triangle BKN$ does not, since the circumcircles of ABC and BKN intersect in two distinct points. It follows that the three

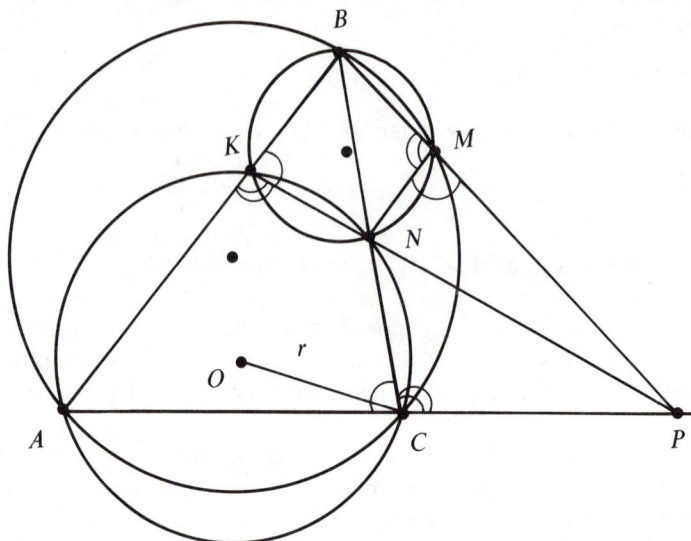

radical axes concur at a point P, called the *radical center*, whose powers with respect to all three circles are equal.† Also, it follows easily that $\angle PMN = \angle BKN = \angle NCA$, and thus $PMNC$ is a cyclic quadrilateral. Then by the power of a point theorem,

(1) $$BM \cdot BP = BN \cdot BC = BO^2 - r^2,$$

(2) $$PM \cdot PB = PN \cdot PK = PO^2 - r^2,$$

where $r = OC$ is the radius of the circle with center O. Subtracting (1) from (2), we get

(3) $$PO^2 - BO^2 = BP(PM - BM) = PM^2 - BM^2,$$

which implies that OM is an altitude of $\triangle OBP$.

1985 / 6. Let $P_1(x) = x$ and define

(1) $$P_{n+1}(x) = P_n(x)\left(P_n(x) + \frac{1}{n}\right) \text{ for } n = 1, 2, \ldots$$

---

†For a full discussion of radical axes and power of a point with respect to a circle, see e.g. *Geometry Revisited* by H. S. M. Coxeter and S. L. Greitzer, NML vol. 19, Sections 2.2, 2.3, especially Theorem 2.31, p. 35. See also Glossary.

From this recursive definition, we see inductively that

(i) P_n is a polynomial of degree 2^{n-1},

(ii) P_n has positive coefficients, is therefore an increasing, convex function for $x \geqslant 0$,

(iii) $P_n(0) = 0$, $P_n(1) \geqslant 1$,

(iv) $P_n(x_1) = x_n$.

Since the condition $x_{n+1} > x_n$ is equivalent to

$$x_n > 1 - \frac{1}{n},$$

we can reformulate the problem as follows: Show that *there is a unique positive real number t such that*

$$1 - \frac{1}{n} < P_n(t) < 1 \qquad\qquad \text{for every } n.$$

Since P_n is continuous and increases from 0 to a value $\geqslant 1$ for $0 \leqslant x \leqslant 1$, there are unique values a_n and b_n such that

$$(2) \qquad a_n < b_n, \qquad P_n(a_n) = 1 - \frac{1}{n}, \qquad P_n(b_n) = 1.$$

Since by definition (1)

$$P_{n+1}(a_n) = \left(1 - \frac{1}{n}\right)\left(1 - \frac{1}{n} + \frac{1}{n}\right) = 1 - \frac{1}{n}$$

and

$$P_{n+1}(a_{n+1}) = 1 - \frac{1}{n+1},$$

we see that

$$(3) \qquad\qquad a_n < a_{n+1}.$$

Also, since $P_{n+1}(b_n) = 1 + \frac{1}{n}$ and $P_{n+1}(b_{n+1}) = 1$,

$$(4) \qquad\qquad b_n > b_{n+1}.$$

And since P_n is convex, the graph of $P_n(x)$ lies below the chord $y = \dfrac{1}{b_n} x$ for $0 \leqslant x \leqslant b_n$, so

$$P_n(x) \leqslant \frac{x}{b_n} \qquad \text{for } 0 \leqslant x \leqslant b_n.$$

In particular, $P_n(a_n) = 1 - \dfrac{1}{n} \leqslant \dfrac{a_n}{b_n}$. From this and the fact that $b_n \leqslant 1$

we find that

$$b_n - \frac{b_n}{n} \leqslant a_n, \qquad b_n - a_n \leqslant \frac{b_n}{n} \leqslant \frac{1}{n} \qquad \text{for all } n.$$

Thus we have two infinite bounded sequences $\{a_n\}$, $\{b_n\}$, the first is increasing, the second decreasing, $a_n < b_n$, and the difference between their n-th members approaches 0 as n increases; we conclude that there is a unique common value t that they approach:

$$a_n < t < b_n \qquad \text{for all } n.$$

This number uniquely satisfies

$$1 - \frac{1}{n} < P_n(t) < 1 \qquad \text{for all } n.$$

Algebra

A/1. Let $\mathbf{v}_k$ be the vector of length 1 forming an angle $2\pi k/n$ with the positive x-axis. We prove a sharpened form of the statement, namely that we can choose a permutation $(b_1, b_2, \ldots, b_n)$ of $(1, 2, \ldots, n)$ such that the vector

$$(1) \qquad \mathbf{S} = b_1\mathbf{v}_1 + b_2\mathbf{v}_2 + \cdots + b_n\mathbf{v}_n = 0.$$

Here we wrote the sum so that the angles (instead of the coefficients) should be in their natural order. The generalization consists of showing that the entire vector can be made $\mathbf{o}$ rather than just the x-component as required in the problem.

If n has at least two distinct prime factors it is the product of two relatively prime integers, $n = pq$, with $p > 1$ and $q > 1$. The construction is based on the following facts: For $r = 1, 2, \ldots, p$,

$$(2) \qquad \mathbf{v}_r + \mathbf{v}_{r+p} + \mathbf{v}_{r+2p} + \cdots + \mathbf{v}_{r+(q-1)p} = \mathbf{o}.$$

The reason for this is that the vectors in each of these sums form a regular q-gon, which is a closed polygon. Similarly, for $s = 1, 2, \ldots, q$,

$$(3) \qquad \mathbf{v}_s + \mathbf{v}_{s+q} + \mathbf{v}_{s+2q} + \cdots + \mathbf{v}_{s+(p-1)q} = \mathbf{o}.$$

Now multiply the r-th equation (2) by r, the s-th equation (3) by $(s-1)p$, and sum the resulting equations over the p values of r and the q values of s. Let b_k be the coefficient of $\mathbf{v}_k$ in the resulting equation. We have to show that every integer from 1 to n occurs once among the b_k. This is a consequence of the Chinese Remainder Theorem.† The above construction says that if $k \equiv r \pmod{p}$ and $k \equiv s \pmod{q}$, then $b_k = r + (s-1)p$. We know that each pair of remainders (r, s) with r and s in the above ranges will occur exactly once as k goes from 1 to n. Also, each integer between 1 and n can be represented in exactly one way in the form $r + (s-1)p$ with r and s in the given ranges. This verifies that the above solution is correct.

We show a figure to illustrate the construction for $n = 15$, $p = 5$, $q = 3$. The first addend at the end of each vector is contributed by the equation of the type (2) and the second, by the equation of type (3). We might note that the construction can be varied considerably.

---

†See Glossary.

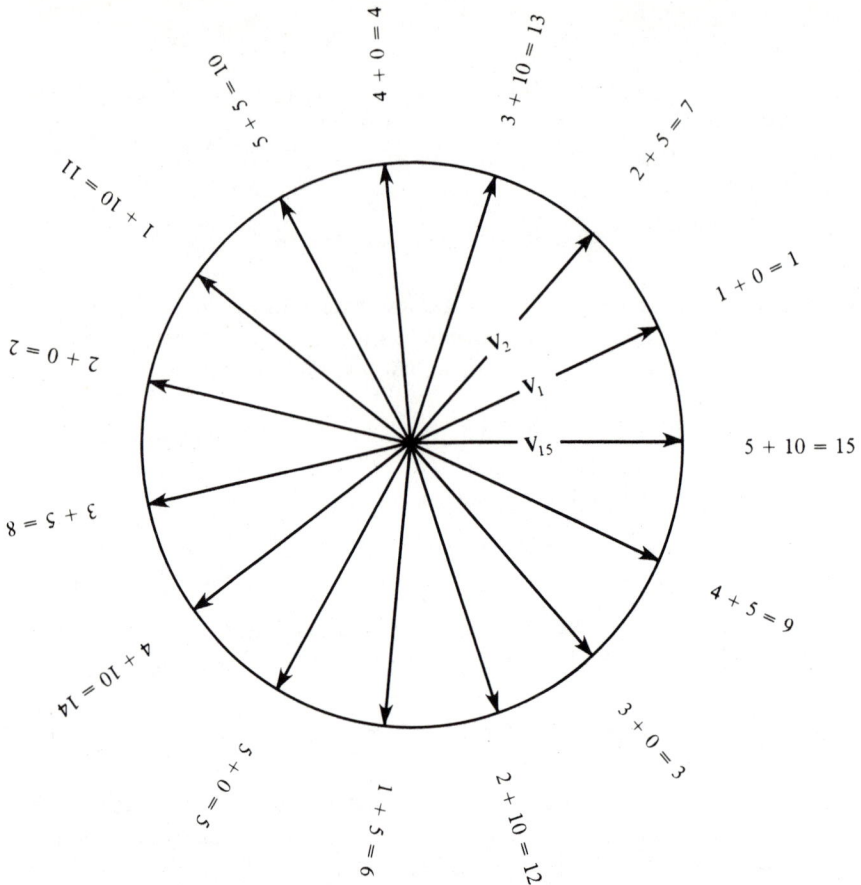

Remark. It is of interest to note that if n is a prime, $n = p$, and (1) holds with integer coefficients, then all the b_i must be equal.

To prove this we regard our vectors as complex numbers. Then $v_k = v_1^k$ by De Moivre's Theorem. Also, $v_1, \ldots, v_p = 1$ are the roots of

$$(4) \qquad\qquad\qquad x^p - 1 = 0.$$

We now need some facts of algebra for which we refer to a college algebra text.†
We need the fact that the second factor in

$$(5) \qquad x^p - 1 = (x - 1)(x^{p-1} + x^{p-2} + \cdots + 1) = (x - 1)P(x)$$

is irreducible. The complex number v_1 is a zero of both polynomials

$$(6) \qquad\qquad\qquad b_p x^{p-1} + b_{p-1} x^{p-2} + \cdots + b_1$$

---

†For example, see G. Birkhoff and S. MacLane, *A Survey of Modern Algebra*, MacMillan, N. Y. (1953).

and

(7) $$x^{p-1} + x^{p-2} + \cdots + 1.$$

Thus the two polynomials (6) and (7) have $x - v_1$ as a common factor. Consequently the greatest common factor with integer coefficients of (6) and (7), given by Euclid's algorithm, is not constant. Since (7) is irreducible the greatest common factor must be a multiple of (7). Since (6) has the same degree as (7), (6) must be just an integer times (7). This completes the proof.

A/2. By the Newton-Gregory interpolation formula† we know that

(1) $$P(x) = \sum_{j=0}^{990} \binom{x - 992}{j} \Delta^j F_{992},$$

where $\Delta^j F_{992}$ are the successive forward differences at 992 of the sequence $F_{992}, F_{993}, \ldots, F_{1982}$. For an *arbitrary* sequence $a_{992}, a_{993}, \ldots, a_{1982}$, these differences are the numbers on the left diagonal of the array

(2)

$$
\begin{array}{cccccc}
a_{992} & a_{993} & a_{994} & a_{995} & \cdots & a_{1982} \\[4pt]
\Delta a_{992} & \Delta a_{993} & \Delta a_{994} & \cdots & \Delta a_{1981} \\[4pt]
\Delta^2 a_{992} & \Delta^2 a_{993} & \cdots & \Delta^2 a_{1980} \\[4pt]
\Delta^3 a_{992} & \cdots & \Delta^3 a_{1979} \\[4pt]
\cdot \quad \cdot & & \cdot \quad \cdot \\[4pt]
\Delta^{990} a_{992}
\end{array}
$$

where $\Delta a_i = a_{i+1} - a_i$, $\Delta^2 a_i = \Delta a_{i+1} - \Delta a_i = (a_{i+2} - a_{i+1}) - (a_{i+1} - a_i) = a_{i+2} - 2a_{i+1} + a_i$, etc.

In the special case where $a_i = F_i$, a major simplification occurs. Since $\Delta F_i = F_{i+1} - F_i = F_{i-1}$, the second row of (2) becomes $F_{991} \, F_{992} \, F_{993} \cdots F_{1980}$. By the same token, the third row of (2) becomes $F_{990} \, F_{991} \cdots F_{1979}$, etc. In particular we have $\Delta^j F_{992} = F_{992-j}$, and therefore (1) can be written in the form

(3) $$P(x) = \sum_{j=0}^{990} \binom{x - 992}{j} F_{992-j}.$$

We are now ready to prove that

(4) $$P(1983) = F_{1983} - 1.$$

Let $Q(x)$ be the polynomial of degree 991 such that $Q(k) = F_k$ for $992 \leqslant k \leqslant 1983$. By the same reasoning used to derive (3), we have

$$Q(x) = \sum_{j=0}^{991} \binom{x - 992}{j} F_{992-j}.$$

†See e.g. E. Whittaker, G. Robinson, *Calculus of Observations*, Dover, N. Y. (1967) pp 10, 16.

Thus $Q(x) = P(x) + \left(\dfrac{x - 992}{991} \right)$.

We now set $x = 1983$ and use the fact that $Q(1983) = F_{1983}$. This gives

$$F_{1983} = P(1983) + 1,$$

which proves (4).

A/3. By letting $n = 0, 1, 2, 3$, successively, we find that $u_0 = 1$, $u_1 = 2$, $u_2 = 2^2$, $u_3 = 2^3$. Consequently, we conjecture that $u_n = 2^n$ for all n; we will establish this result by induction. We assume that $u_k = 2^k$ for $k = 0, 1, 2, \ldots, n - 1$. Then from the given relation and the induction hypothesis,

$$(1) \qquad u_n^2 - u_n = \sum_{r=1}^{n} \binom{n + r}{r} u_{n-r} = \sum_{r=1}^{n} \binom{n + r}{r} 2^{n-r}.$$

If it were known that

$$(2) \qquad \sum_{r=1}^{n} \binom{n + r}{r} 2^{n-r} = 2^{2n} - 2^n,$$

it would then follow from (1) that

$$(u_n - 2^n)(u_n + 2^n - 1) = 0;$$

and since $u_n > 0$, $u_n = 2^n$ which would complete the induction and establish $u_n = 2^n$ as the unique solution of the problem.

Now (2) is easily shown to be equivalent to

$$(3) \qquad 2^n = \sum_{r=0}^{n} \binom{n + r}{r} 2^{n-r}.$$

Although (3) is a known binomial identity, we give several proofs including generalizations.

P_1: Denote the right side of (3) by a_n. Using the identity

$$\binom{n + r}{r} = \binom{n - 1 + r}{r - 1} + \binom{n - 1 + r}{r}$$

we obtain

$$a_n = \frac{1}{2} \sum_{r=0}^{n} \binom{n - 1 + r}{r - 1} 2^{n-(r-1)} + 2 \sum_{r=0}^{n} \binom{n - 1 + r}{r} 2^{n-1-r}$$

$$= \frac{1}{2} \sum_{s=0}^{n-1} \binom{n + s}{s} 2^{n-s} + 2a_{n-1} + \binom{2n - 1}{n},$$

where we have made the substitution $r - 1 = s$ in the first term. Thus,

$$a_n = \frac{1}{2} \left(a_n - \binom{2n}{n} \right) + 2a_{n-1} + \binom{2n - 1}{n} = \frac{1}{2} a_n + 2a_{n-1},$$

which reduces to

$$a_n = 4a_{n-1}.$$

Since $a_0 = 1$, we get $a_n = 4^n = 2^{2n}$ by induction.

Our next two proofs are probabilistic.

P_2: The identity (3), which we re-write in the form

$$\sum_{r=0}^{n} \binom{n+r}{n} 2^{-n-r} = 1,$$

corresponds to the special case $x = 1/2$ in the more general identity

(4) $$\sum_{r=0}^{n} \binom{n+r}{n}\{(1-x)^{n+1}x^r + x^{n+1}(1-x)^r\} = 1.$$

To derive (4), assume two teams A and B are playing a World Championship Series. The first team to win $n+1$ games (there are no ties) will be the World Champion. Let x be the probability that team A beats team B in any game and then $1-x$ is the probability that B beats A in any game. In a $n+1+r$ game series where r can be $0,1,2,\ldots,n$, A can only win the series by winning the last game and any n games of the first $n+r$ games played. Consequently, the probability that A wins the series is given by

$$P(A) = \sum_{r=0}^{n} \binom{n+r}{n} x^{n+1}(1-x)^r.$$

Similarly, the probability that B wins the series is

$$P(B) = \sum_{r=0}^{n} \binom{n+r}{n}(1-x)^{n+1}x^r.$$

Finally, we must have $P(A) + P(B) = 1$ which is identity (4).

In a similar fashion†, (4) can be generalized to

$$S(x,y,z,m,n,r) + S(y,z,x,n,r,m) + S(z,x,y,r,m,n) = 1$$

where

$$S(x,y,z,m,n,r) = x^{m+1}\sum_{j=0}^{n}\sum_{k=0}^{r} y^j z^k (j+k+m)!/j!k!m!$$

and $x + y + z = 1$.

P_3: For our last proof, we consider a famous problem in probability theory, known as *Banach's matchbox problem*. It is usually formulated as

†For proofs and extensions to any number of variables, see Problem 85-10, SIAM Review 28 (1986) 243–244.

follows. An absent-minded professor carries 2 matchboxes, labelled A and B, in his pockets. Each originally contains n matches. When the professor wants to smoke, he chooses a matchbox at random, selects a match from it, and lights his pipe. One day he chooses matchbox A, but finds that it is empty. What is the probability that matchbox B contains exactly r matches $(0 \leqslant r \leqslant n)$?

To solve this problem we suppose that each box contains *infinitely* many matches! We then allow the professor to select a total of $n + r$ matches from the two boxes. The problem now becomes: what is the probability that in this process, n matches are selected from A, and r matches are selected from B? This probability is clearly $\binom{n+r}{r} / 2^{n+r}$. The sum of these probabilities from $r = 0$ to $r = n$ is 1, so

$$\sum_{r=0}^{n} \binom{n+r}{r} \frac{1}{2^{n+r}} = 1,$$

which proves (3).

Remark. In the matchbox problem, suppose the boxes are not chosen at random, but rather that A is chosen with probability p, while B is chosen with probability $1 - p$, where $0 \leqslant p \leqslant 1$. Then the desired probability is no longer $\binom{n+r}{r} / 2^{n+r}$, but rather $\binom{n+r}{r} p^n (1 - p)^r$. Since these probabilities sum to 1, we obtain the more general identity

$$(5) \qquad \sum_{r=0}^{n} \binom{n+r}{r} p^n (1 - p)^r = 1.$$

To see how (4) follows from (5), interchange p and $1 - p$, also n and r, in (5):

$$(5') \qquad \sum_{r=0}^{n} \binom{n+r}{n} (1 - p)^r p^n = 1.$$

Multiply (5) by p, (5') by $(1 - p)$, and add. The resulting identity is (4) with $x = 1 - p$. Question: Can (5) be deduced from (4)?

A/4. Define

$$P_n = \frac{a}{(bc - a^2)^n} + \frac{b}{(ca - b^2)^n} + \frac{c}{(ab - c^2)^n}, \qquad n = 1, 2,$$

and

$$Q = \frac{1}{bc - a^2} + \frac{1}{ca - b^2} + \frac{1}{ab - c^2}.$$

Multiplying Q and P_1 we find after some algebraic simplification that $QP_1 = P_2$, an identity valid for all real numbers a, b, c for which the denominators in Q are not zero. If $Q = 0$ for certain real numbers a, b, c, then also $P_2 = 0$, as was to be proved.

We note that an additional consequence of this proof is that if $P_1 = 0$ then $P_2 = 0$. Furthermore, the results hold not only for real numbers, but more generally for complex numbers.

A/5. The key to the solution of this problem is the identity

(1) $$\tan A \tan B \tan C = \tan A + \tan B + \tan C$$

for any 3 angles A, B, C with sum $180°$. To prove this we note that

$$\tan C = \tan(180° - A - B) = -\tan(A + B) = \frac{\tan A + \tan B}{\tan A \tan B - 1},$$

which proves (1).

Now denote the common value of the two sides of (1) by a, and set

$$b = \tan A \tan B + \tan A \tan C + \tan B \tan C.$$

Then

(2) $$(x - \tan A)(x - \tan B)(x - \tan C) = x^3 - ax^2 + bx - a.$$

Hence, if we denote the remaining zero of the given polynomial

$$x^4 - px^3 + qx^2 - rx + s$$

by R, we have

$$(x^3 - ax^2 + bx - a)(x - R)$$
$$= x^4 - (a + R)x^3 + (b + aR)x^2 - (a + bR)x + aR$$
$$= x^4 - px^3 + qx^2 - rx + s.$$

Equating coefficients of $x^3, x^2, x, 1$ in both expressions, we obtain the following four equations:

(i) $a + R = p,$ (ii) $b + aR = q,$

(iii) $a + bR = r,$ (iv) $aR = s.$

Our task is to eliminate a and b, and solve for R. Subtracting (iv) from (ii), we obtain

$$b = q - s.$$

Subtracting (iii) from (i), we get

$$(1 - b)R = p - r.$$

If $b \ne 1$, we have

$$R = \frac{p - r}{1 - b} = \frac{p - r}{1 - q + s}.$$

If $b = 1$, the right hand side of (2) becomes $x^3 - ax^2 + x - a = (x - a)(x^2 + 1)$. Since this polynomial has only one real zero (while the problem asserts that there are at least three), this case cannot occur. Thus in all cases $R = (p - r)/(1 - q + s)$.

A/6 First solution. The hypothesis that $S_1 = S_2 = \cdots = S_{n+1}$ is unnecessarily strong. We need it for $n = 1$ and $n = 2$, but for $n \geqslant 3$ we need only the weaker hypothesis that $S_2 = S_3 = S_4$.

Case 1. $n = 1$. The condition $S_1 = S_2$ is $x_1 = x_1^2$. This clearly implies that $x_1 = 0$ or 1.

Case 2. $n = 2$. The condition $S_1 = S_2 = S_3$ is $x_1 + x_2 = x_1^2 + x_2^2 = x_1^3 + x_2^3$. It follows from this that $(x_1 + x_2)(x_1^3 + x_2^3) = (x_1^2 + x_2^2)^2$, i.e.

$$x_1^4 + x_1^3 x_2 + x_1 x_2^3 + x_2^4 = x_1^4 + 2x_1^2 x_2^2 + x_2^4.$$

Subtracting the terms x_1^4 and x_2^4 from both sides, we obtain

$$(1) \qquad\qquad x_1 x_2 (x_1^2 + x_2^2) = 2x_1^2 x_2^2.$$

If $x_1 = 0$, the equation $x_1 + x_2 = x_1^2 + x_2^2$ becomes $x_2 = x_2^2$, which implies that $x_2 = 0$ or 1. So if $x_1 = 0$, the proof is complete. The same applies if $x_2 = 0$.

If $x_1 x_2 \neq 0$, we may divide both sides of (1) by $x_1 x_2$, getting $x_1^2 + x_2^2 = 2x_1 x_2$. Thus $(x_1 - x_2)^2 = x_1^2 + x_2^2 - 2x_1 x_2 = 0$, and hence $x_1 = x_2$. The equation $x_1 + x_2 = x_1^2 + x_2^2$ now yields $2x_1 = 2x_1^2$, and since $x_1 \neq 0$, this implies that $x_1 = x_2 = 1$.

Case 3. $n \geqslant 3$. Consider the sum of squares

$$(2) \qquad \sum_{i=1}^{n} (x_i - x_i^2)^2 = \sum_{i=1}^{n} (x_i^2 - 2x_i^3 + x_i^4) = S_2 - 2S_3 + S_4.$$

Since $S_2 = S_3 = S_4$, the right side of (2) is 0. Therefore, so is its left side, which can be only if

$$x_1 - x_1^2 = x_2 - x_2^2 = \cdots = x_n - x_n^2 = 0.$$

This implies that each x_i is either 0 or 1.

Note that we could just as easily deduce the conclusion from the hypothesis $S_{2i} = S_{i+j} = S_{2j}$ for some pair i, j. Also, if the x_i are required to be positive rather than real, then the conclusion follows from the hypothesis $S_i = S_j = S_k$ for some distinct i, j, k. This can be proved using Hölder's inequality† and we leave it as an exercise for the reader.

Second solution. In this solution we will use the full hypothesis that $S_1 = S_2 = \cdots = S_{n+1}$. By way of compensation, the proof works not only for real numbers x_i, but for elements x_i from any integral domain‡, for example the complex numbers.

---

†See Glossary.
‡G. Birkhoff and S. McLane, *A Survey of Modern Algebra*, MacMillan, N. Y. (1953), Ch. 1.

If any of the numbers x_i are 0 or 1, we remove them and note that the remaining x_i still satisfy the hypothesis.

Thus we may suppose from now on that no element x_i is 0 or 1. We will proceed to derive a contradiction. Let

(3)
$$P(x) = x^n + a_1 x^{n-1} + a_2 x^{n-2} + \cdots + a_n$$
$$= (x - x_1)(x - x_2)\ldots(x - x_n)$$

be the monic polynomial whose zeros are the numbers $x_1, x_2, \ldots, x_n$. We set $x = x_i$; then

$$x_i^n + a_1 x_i^{n-1} + a_2 x_i^{n-2} + \cdots + a_n = 0 \qquad \text{for } 1 \leqslant i \leqslant n.$$

Summing these equations over all i, we obtain

$$S_n + a_1 S_{n-1} + a_2 S_{n-2} + \cdots + n a_n = 0.$$

Thus if S denotes the common value of $S_1, S_2, \ldots, S_{n+1}$, we have

(4)
$$S(1 + a_1 + a_2 + \cdots + a_{n-1}) + n a_n = 0.$$

Now multiply (3) by x and then substitute $x = x_i$. Again the right side vanishes, so

$$x_i P(x_i) = x_i^{n+1} + a_1 x_i^n + a_2 x_i^{n-1} + \cdots + a_n x_i = 0 \quad \text{for } 1 \leqslant i \leqslant n.$$

Summing over all i, we get

$$S_{n+1} + a_1 S_n + a_2 S_{n-1} + \cdots + a_n S_1 = 0,$$

and hence

(5)
$$S(1 + a_1 + a_2 + \ldots a_{n-1}) + S a_n = 0.$$

Comparison of (4) and (5) gives

(6)
$$S a_n = n a_n.$$

From (3) we see that $a_n = (-1)^n x_1 x_2 \ldots x_n$, and since we are treating the case where no x_i is 0, we have $a_n \neq 0$. Therefore (6) implies that

(7)
$$S = n.$$

Now consider the quantities

$$y_i = 1 - x_i.$$

Denote their power sums by T_k: $T_k = \Sigma y_i^k = \Sigma(1 - x_i)^k$. Thus

$$T_k = n - \binom{k}{1} S_1 + \binom{k}{2} S_2 - \binom{k}{3} S_3 + \cdots + (-1)^k S_k$$
$$= n + \left[-\binom{k}{1} + \binom{k}{2} - \binom{k}{3} + \cdots + (-1)^k \right] S.$$

By the binomial theorem,

$$0 = (1 - 1)^k = 1 - \binom{k}{1} + \binom{k}{2} - \binom{k}{3} + \cdots + (-1)^k.$$

Therefore

$$-\binom{k}{1} + \binom{k}{2} - \binom{k}{3} + \cdots + (-1)^k = -1,$$

and so

$$T_k = n - S.$$

By (7), we conclude that

(8) $$T_k = 0, \qquad k = 1, 2, \ldots, n + 1.$$

This shows that the set $\{y_1, y_2, \ldots, y_n\}$ satisfies the conditions of the problem. Since no x_i is 1, none of the $y_i = 0$. Therefore (7) is satisfied by the set $\{y_1, y_2, \ldots, y_n\}$:

$$T_k = n,$$

in contradiction to (8).

Remark: Another way to complete the argument, starting from (7), is to appeal to a classical result of Newton's†:

The elementary symmetric functions $a_1, a_2, \ldots, a_n$ of $\{x_1, x_2, \ldots, x_n\}$ appearing in (3) can be expressed in terms of the power sums $S_1, S_2, \ldots, S_n$. Since $\{x_1, x_2, \ldots, x_n\}$ is the set of zeros of the polynomial (3), it follows that $S_1, S_2, \ldots, S_n$ determines $\{x_1, x_2, \ldots, x_n\}$ uniquely. According to (7), $S_1 = S_2 = \cdots = S_n = n$; this is satisfied by $x_1 = x_2 = \cdots = x_n = 1$, and therefore by no other set.

A/7. For motivation, we first consider the simpler problem where the cubic polynomial identity is replaced by

(1) $$a^2 + b^2 + c^2 = bc + ca + ab.$$

It follows that we should be able to rewrite (1) in such a form that the implications $b - c = c - a = a - b$ are fairly evident. This is indeed valid, since (1) is equivalent to $(b - c)^2 + (c - a)^2 + (a - b)^2 = 0$.

---

†The Newton formulae are

$$S_1 + a_1 = 0, \quad S_2 + a_1 S_1 + 2a_2 = 0, \ldots, S_{n-1} + a_1 S_{n-2} + \cdots + (n-1)a_{n-1} = 0,$$

and

$$S_r + a_1 S_{r-1} + \cdots + a_n S_{r-n} = 0 \text{ for } r \geqslant n,$$

where $a_1 = -\Sigma x_i$, $a_2 = \Sigma x_i x_j$, $a_3 = -\Sigma x_i x_j x_k, \ldots$ see eq. (3). For several different proofs, see M. S. Klamkin and D. J. Newman, "Uniqueness Theorems for Power Equations", Elem. der Mathematik 25/6 (1970) pp. 130–134, and references therein; also treated there is the problem: For which values of w is the implication $S_1 = S_2 = \cdots = S_n = w \Rightarrow w = S_{n+1} = S_{n+2} = \cdots$ valid? As before, x_i and w may be real or complex. Another related but more difficult problem (due to D. J. Newman) is to determine S_{n+1}, given that $S_k = k$ for $k = 1, 2, \ldots, n$. For a solution, see Amer. Math. Monthly 82 (1975) 764–765.

For the given problem, we now expect to be able also to express the given identity as a sum of non-negative terms which would imply that $a = b = c$. Since the given identity is a cubic and $(b - c)^2$, etc. are quadratics, we will have to multiply the latter terms by some non-negative linear expressions in a, b, c.

For a triangle, $b + c - a$, $c + a - b$, and $a + b - c$ are all > 0. After some trial and error, we find that the given identity can be expressed as sum of the non-negative terms

$$(a + b - c)(b - c)^2 + (b + c - a)(c - a)^2 + (c + a - b)(a - b)^2 = 0.$$

Hence, $a = b = c$.

For the second part, let $b = 3$, $c = 1$, and then the given relation reduces to

$$F(a) = 4(a + 2) + (4 - a)(a - 1)^2 + (a - 2)(a - 3)^2 = 0.$$

Then since $F(0) = -6$ and $F(1) = 8$, there is at least one value of a in $(0, 1)$ for which $F(a) = 0$. But then $c + a < 2 < b$.

A/8. The sequences a_n and b_n are strictly increasing. Consequently, if $a_n = b_m$ and $a_{n'} = b_{m'}$ with $n < n'$, then $m < m'$. Each of the recurrence equations can be solved explicitly to give

$$a_{n+1}/n! = a_1 + (1/1! + 1/2! + \cdots + 1/n!),$$

$$b_{m+1}/m! = b_1 - (1/1! + 1/2! + \cdots 1/m!).$$

To show that there cannot be an infinite number of pairs (n, m) such that $a_{n+1} = b_{m+1}$, we assume there are and obtain a contradiction. For each of these pairs, we obtain (by subtracting $m!/n!$ times the second equation from the first)

$$(1) \quad \frac{a_{n+1}}{n!} - \frac{b_{m+1}}{n!} = 0 = a_1 + \left(\frac{1}{1!} + \frac{1}{2!} + \cdots + \frac{1}{n!} \right)$$
$$- \left(b_1 - \frac{1}{1!} - \frac{1}{2!} - \cdots - \frac{1}{m!} \right) \frac{m!}{n!}.$$

If there are infinitely many pairs such that $a_{n+1} = b_{m+1}$, then at least one of the cases $n > m$, $n = m$, or $n < m$ holds for infinitely many, and therefore for arbitrarily large m and n. Recall that $1 + 1/1! + 1/2! + \cdots = e$. Then for $n > m$ and for $n = m$, the right side of (1) tends to $a_1 + e - 1$ and to $a_1 - b_1 + 2(e - 1)$, respectively. Since a_1, b_1 are natural numbers, neither expression vanishes, and so at most finitely many pairs have $n > m$ or $n = m$. And in the case $n < m$, the right side tends to $-\infty$ or ∞ depending on whether $b_1 > e - 1$ or $b_1 < e - 1$. We conclude that our sequences have at most finitely many common members.

Remark. The above proof rests on the assumption that both a_1 and b_1 are positive integers. However, the conclusion holds without these assumptions. To show this we need to make more precise the difference between e and the finite sections of the factorial series:

$$e = 1 + \frac{1}{1!} + \frac{1}{2!} + \cdots + \frac{1}{n!} + \frac{d_n}{(n+1)!},$$

where

$$\lim_{n \to \infty} d_n = 1.$$

As before we write

$$0 = \frac{a_{n+1}}{n!} - \frac{b_{m+1}}{n!} = (a_1 + e - 1) - \frac{d_n}{(n+1)!}$$

$$-(b_1 - e + 1)\frac{m!}{n!} - \frac{d_m}{(m+1)n!}.$$

The case $n > m$: If

$$a_1 + e - 1 \neq 0,$$

the first term, for n large enough, is much larger than the other terms, and so the sum of the four terms on the right cannot be zero.

If

$$a_1 + e - 1 = 0 \quad \text{and} \quad b_1 - e + 1 \neq 0,$$

then for n large enough the third term is much larger than the remaining two, and so for large n the sum on the right cannot be zero.

If both

$$a_1 + e - 1 = 0, \quad b_1 - e + 1 = 0,$$

then since the remaining two nonzero terms on the right are both negative, their sum cannot be zero for any n and m.

The two other cases can be handled similarly.

A/9 First solution. We show that all the terms of the sequence are equal. Since $a_1 = 1$, this will imply that $a_i = 1$ for all i.

By the triangle inequality,

$$|a_2 - a_1| = |a_2 - a_n + a_n - a_1| \leqslant |a_2 - a_n| + |a_1 - a_n|;$$

and by hypothesis,

$$|a_2 - a_1| \leqslant \frac{2 \cdot 2n}{2^2 + n^2} + \frac{1 \cdot 2n}{1^2 + n^2} < \frac{6n}{n^2} = \frac{6}{n}.$$

By taking n sufficiently large, we can make $6/n$ arbitrarily small; therefore $|a_2 - a_1|$ must be zero, and $a_2 = a_1 = 1$. The same argument may be applied to $|a_j - a_k|$ for any pair j, k. We conclude that $a_1 = a_2 = \cdots = 1$.

Second solution. We will show that all the terms of the sequence are 1's. Hold m fixed and let $n \to \infty$. Then

$$\lim_{n \to \infty} \frac{2mn}{m^2 + n^2} = 0.$$

Since $|a_m - a_n| \leqslant 2mn/(m^2 + n^2)$, this implies that $\lim_{n \to \infty} |a_m - a_n| = 0$, i.e., $\lim_{n \to \infty} a_n = a_m$. Since this holds for every m, and a sequence can only have one limit, all the a_m's are equal. So they are all equal to $a_1 = 1$.

A/10. Three obvious solutions of

$$(1) \qquad\qquad f(x + y)f(x - y) = (f(x)f(y))^2$$

are $f(x) \equiv 0$, 1 or -1.

Setting $y = 0$, we get $(f(x))^2 = (f(x))^2(f(0))^2$, so that, if $f(x) \neq 0$ for some value of x, then $f(0) = 1$ or -1. Since f satisfies (1) if and only if $-f$ does, it suffices to consider the case $f(0) = 1$.

If we put $x = 0$, we get $f(y)f(-y) = (f(y))^2$. If $f(y) \neq 0$, we can divide and get

$$(2) \qquad\qquad f(-y) = f(y).$$

Equation (2) still holds if both $f(y)$ and $f(-y)$ are 0, so we have shown that f is an even function.

Putting $x = y$, we get

$$(3) \qquad\qquad f(2x) = (f(x))^4;$$

so

$$f(x) = 0 \text{ implies } f(x/2) = 0.$$

Thus if f vanishes anywhere, then it vanishes on a set of points approaching 0. Since $f(0) = 1$ and f is continuous, that cannot happen. Consequently

$$f(x) > 0$$

everywhere.

We claim now that for all natural numbers n,

$$(4)_n \qquad\qquad f(nx) = (f(x))^{n^2},$$

For $n = 1$, $(4)_1$ holds trivially; for $n = 2$, $(4)_2$ is equation (3). We prove $(4)_n$ by induction, setting $y = kx$ in (1):

$$f((k + 1)x)f((k - 1)x) = (f(x)f(kx))^2 = (f(x))^2(f(kx))^2.$$

We use $(4)_k$ and $(4)_{k-1}$ to obtain $(4)_{k+1}$. This completes the induction.

Setting $x = 1/n$ in $(4)_n$ we get

$$f(1) = \left(f\left(\frac{1}{n}\right)\right)^{n^2}, \qquad \text{and thence} \qquad f\left(\frac{1}{n}\right) = (f(1))^{1/n^2};$$

and using (4) again, we find $f(m \cdot 1/n) = (f(1/n))^{m^2}$, so

(5)
$$f\left(\frac{m}{n}\right) = (f(1))^{m^2/n^2}.$$

Thus for all positive rational values of x,

(6)
$$f(x) = (f(1))^{x^2}.$$

By continuity (6) holds for positive irrational values of x also. To cover negative values of x we use that both sides of (6) are even. Thus the nonzero solutions of our problem are the functions of the form $\pm a^{x^2}$, $a > 0$.

Note. Setting $y = x$ in (1) and observing that $f(0) = \pm 1$, say $+1$, we obtained equation

(3)
$$f(2x) = f^4(x).$$

Thus all functions that satisfy (1) satisfy (3). It is noteworthy that all functions f which satisfy (3) *and are twice differentiable* at $x = 0$ are of form

$$f(x) = a^{x^2};$$

therefore they satisfy (1). By "twice differentiable at $x = 0$" we mean that for x near 0,

(7)
$$f(x) = 1 + bx + c'x^2 + \varepsilon'x^2,$$

where $\varepsilon'(x) \to 0$ as $x \to 0$.

PROOF: It is convenient to introduce as new function

$$l(x) = \log f(x);$$

this is possible since we have shown above that $f(x) \neq 0$. Taking the log of (3) gives the following functional equation for l:

(3)′
$$l(2x) = 4l(x), \qquad l(0) = 0,$$

a linear equation. It follows from (7) that l is twice differentiable at the origin:

(7)′
$$l(x) = bx + cx^2 + \varepsilon x^2,$$

where $\varepsilon(x) \to 0$ as $x \to 0$.
    Replacing x by $x/2, x/4, \ldots$, in (3)′ gives, for any integer n,

$$l(x) = 4l(x/2) = \cdots = 4^n l(x/2^n).$$

Using (7)′ we get

$$l(x) = 4^n \left(\frac{bx}{2^n} + (c + \varepsilon(x/2^n))\frac{x^2}{4^n} \right)$$

$$= 2^n bx + cx^2 + x^2 \varepsilon(x/2^n).$$

Letting $n \to \infty$, we deduce from this that $b = 0$ and

$$l(x) = cx^2.$$

So

$$f(x) = e^{cx^2} = a^{x^2}.$$

Condition (7) is necessary for deriving (6); for set

(8) $$l(x) = x^p$$

into (3)′;

$$l(2x) = (2x)^p = 2^p x^p = 4l(x) = 4x^p.$$

This is satisfied when

$$2^p = 4.$$

The set of p satisfying this relation is

(9) $$p = 2 + \frac{2\pi i n}{\log 2}.$$

The function (8) is twice differentiable at $x = 0$ only if $n = 0$. The function (8) is a complex valued solution; but since equation (3)′ is linear,

$$l(x) = x^p + x^{\bar{p}},$$

(where $\bar{p}$ is the complex conjugate of p) is also a solution of (3)′, and is real valued.

For a thorough treatment of the Schröder-Koenigs functional equation $\phi(f(x)) = s\phi(x)$, for which (3)′ is a special case, see M. Kuczma, *Functional Equations in a Single Variable*, Polish Scientific Publishers, Warsaw, 1968, Chap. VI.

Number Theory

N.T./1. More generally† we will show by induction on n that for any fixed positive integer m there exists a set of n consecutive positive integers each of which is divisible by a number of the form a^m, where a is some integer greater than 1.

For $n = 1$, clearly a^m satisfies the conditions. Assume that for $n = k$, each of the k consecutive numbers $N_1, N_2, \ldots, N_k$ is divisible by an m-th power > 1. Thus N_i is divisible by a_i^m $(a_i > 1)$ for $i = 1, 2, \ldots, k$. Let $P = (a_1 a_2 \cdots a_k)^m$. We now define $N = N_{k+1}\{(P + 1)^m - 1\}$, where $N_{k+1} = N_k + 1$. Then $N + N_1, N + N_2, \ldots, N + N_{k+1}$ are $k + 1$ consecutive numbers divisible by $a_1^m, a_2^m, \ldots, a_k^m$, $(P + 1)^m$, respectively. Hence the desired result is valid by induction.

N.T./2. Define $a_k = x_k x_{k+1} x_{k+2} x_{k+3}$ for $k = 1, 2, \ldots, n$, where $x_{n+i} = x_i$. By hypothesis $a_k = \pm 1$ and $a_1 + a_2 + \cdots + a_n = 0$. It follows that n must be even, say $n = 2m$, with m of the a_i's equal to $+1$, and the other m equal to -1. Hence $a_1 a_2 \cdots a_n = (-1)^m$. But each x_i

---

†The result here was proposed by E. P. Starke as problem #106 in Math. Mag. 25 (1952), 221. The solution here is by S. B. Akers, Jr. and is one of the three published ones.

appears four times in the product $a_1a_2 \cdots a_n$, and hence this product equals $+1$. Thus m is even, say $m = 2p$, so that $n = 4p$.

(For an alternative solution, see p. 142.)

N.T./3. We first consider the factorization
$$x^5 - 1 = (x - 1)(x^4 + x^3 + x^2 + x + 1),$$
where $x = 5^{397}$. The difficulty now is in factoring $x^4 + x^3 + x^2 + x + 1$. We can verify that

(1) $\qquad x^4 + x^3 + x^2 + x + 1 = (x^2 + 3x + 1)^2 - 5x(x + 1)^2.$

Since $x = 5^{397}$, the right hand side of (1) is the difference of two squares and can be factored. It is easy to verify that each of the three factors of $5^{1985} - 1$ exceeds 5^{100}.

N.T./4. (a) If $(m, n) = 1$, then there are positive integers r and s such that $sn = rm + 1$. Now place the boxes in a row so that no box has fewer balls than the first one. Put one ball in each box starting from the first box, continuing through all boxes, then back to the first again and so on in cyclical fashion. The operation has been used s times with $rm + 1$ balls added to the boxes. Thus each box has received r balls, except the first box which got $r + 1$ balls. Repeat this use of the operation, starting always with n of the boxes lined up so that no box has fewer balls than the first one. Thus we can obtain an equal number of balls in each of the boxes.

(b) The case $(m, n) = d > 1$. If we start with $m + 1$ balls in the m boxes, then after the t-th operation there are $m + 1 + tn$ balls in the boxes. However, this number is never a multiple of m, since it is not divisible by d. Thus an equal distribution cannot be achieved.

N.T./5 (i). Denote the five distinct roots of $x^5 - 1 = 0$ by 1, $\omega, \omega^2, \omega^3, \omega^4$. We use the result that if k is a positive integer, then
$$1 + \omega^k + \omega^{2k} + \omega^{3k} + \omega^{4k} = \begin{cases} 5 & \text{if } k \text{ is a multiple of 5} \\ 0 & \text{otherwise.} \end{cases}$$

The first result is obvious, and the second follows by substitution of $x = \omega^k$ in the identity $x^5 - 1 = (x - 1)(1 + x + x^2 + x^3 + x^4)$.

Now replace x successively by 1, $\omega, \omega^2, \omega^3, \omega^4$ in the identity

(1) $\quad x^2(1 + x + x^2 + x^3 + x^4)^{496} = x^2(a_0 + a_1x + a_2x^2 + a_3x^3 + \cdots)$

(we multiplied the given identity by x^2 in order that $a_3, a_8, \ldots, a_{5n+3}$ be coefficients of fifth powers of x), add the resulting five equations, and divide by 5 to get
$$5^{495} = a_3 + a_8 + a_{13} + \cdots + a_{1983}.$$

This shows that any common divisor of $a_3, a_8, \ldots, a_{1983}$ is a power of 5. By direct observation we see from the given equation that $a_{1983} = 496$, which is not divisible by 5. It follows that the greatest common divisor we seek is 1.

(ii) A. Setting $x = 1$ in the given equation gives

$$5^{496} > a_{992},$$

since $a_i \geqslant 0$. To show that $10^{347} > a_{992}$, it suffices to show that

$$10^{347} > 5^{496}, \quad \text{or that} \quad 347 > 496(1 - \log 2).$$

An easy way to verify the last inequality is to recall that $.3010 < \log 2 < .3011$. Then

$$496(1 - \log 2) < 496(1 - .3010) = 346.704.$$

Alternatively, we can use the fact that $2^{10} = 1024 > 10^3$, multiply our inequality by 2^{496} and raise it to the tenth power. It reduces to

$$2^{4960} > 10^{1490} \quad \text{or} \quad 1.024^{496} > 100 \quad \text{or} \quad (1.024)^{124} > \sqrt{10} \,.$$

In the binomial expansion

$$(1 + .024)^{124} = 1 + 124(.024) + \cdots,$$

the sum of the first two terms is already $> \sqrt{10}$.

(ii) B. It follows inductively that the coefficients $a_0, a_1, \ldots$ are unimodal and symmetric, a_{992} being the largest. Thus $1985\, a_{992} > 5^{496}$. To show that $a_{992} > 10^{340}$, it suffices to establish that

(2) $\quad 5^{496} > 2000 \cdot 10^{340} \quad$ or $\quad 496 - 343 > 497 \log 2 \quad$ or $\quad 153 > 149.647$.

Alternatively, we can use the fact that $5^4 > 2^9$ to conclude first that $(5^4)^{39} > (2^9)^{39}$ or $5^{156} > 2^{351}$; then, since $2^{351} = 2^{11} \cdot 2^{340} > 2 \cdot 10^3 \cdot 2^{340}$, we have that $5^{156} > 2000 \cdot 2^{340}$, which is equivalent to (2).

Remark. A still sharper bound for a_{992} is obtained if we apply to the identity

$$x^3(1 + x + x^2 + x^3 + x^4)^{496} = x^3\big(a_0 + a_1 x + a_2 x^2 + a_3 x^3 + \cdots \big)$$

the same technique used in (i) above to get

$$5^{495} = a_2 + a_7 + a_{12} + \cdots + a_{992} + \cdots + a_{1982}.$$

This gives $5^{495} > a_{992}$. We now prove that

$$10^{347} > 5^{495} \quad \text{or equivalently, that} \quad 2^{347} > 5^{148},$$

to establish part of the inequality in (ii). From $2^7 > 5^3$ we get $(2^7)^{49} > (5^3)^{49}$ or $2^{343} > 5^{147}$, and the result follows from this.

N.T./6. For $n = 0$, the given equation yields $a_1 = k$, an integer. With some algebraic effort we can solve the given equation for a_n in terms of

a_{n+1}. We find

$$a_n = k(a_{n+1} + 1) + (k + 1)a_{n+1} - 2\sqrt{k(k + 1)a_{n+1}(a_{n+1} + 1)}.$$

We now add this equation to the given equation with n replaced by $n + 1$ and obtain

$$a_{n+2} + a_n = 2k(a_{n+1} + 1) + 2(k + 1)a_{n+1},$$

or

$$a_{n+2} = (4k + 2)a_{n+1} - a_n + 2k.$$

From this recursion with integer coefficients and from the fact that a_0, a_1 are integers, it now follows inductively that a_n is a positive integer for $n = 1, 2, \ldots$. For extensions of this result, see Math. Mag. 42 (1969), pp. 111–113.

Plane Geometry

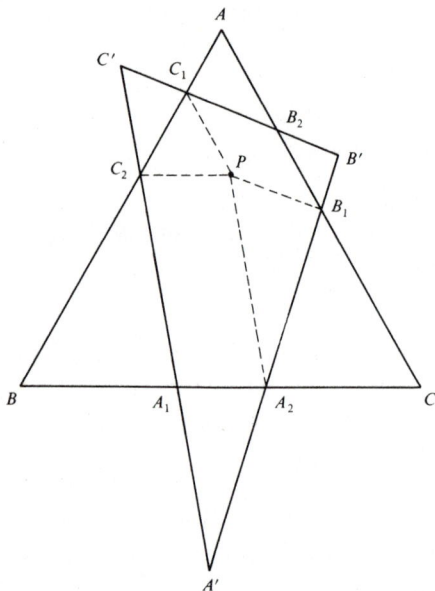

P.G./1. Let the triangle formed by lines B_2C_1, C_2A_1, and A_2B_1 be $A'B'C'$ as shown above. Now consider parallelogram $C_2A_1A_2P$. Then $\triangle C_1C_2P$ is equilateral. Also, since C_1P is equal and $\parallel$ to B_2B_1, $B_1B_2C_1P$ is a parallelogram. Therefore $B_2C_1 : C_2A_1 : A_2B_1 = B_1P : PA_2 : A_2B_1$. Finally $\triangle A_2B_1P \sim \triangle A'B'C'$ giving the desired result.

P.G./2. We choose a rectangular coordinate system with l as the x-axis and with the y-axis going through the center $(0, d)$ of the given circle of radius $r < d$. The center $(h, 0)$ of a circle externally tangent to the given circle must have a radius, say s, which satisfies

$$(1) \qquad\qquad h^2 + d^2 = (r + s)^2$$

(see figure). The points M and N then have coordinates $(h - s, 0)$ and $(h + s, 0)$, respectively.

By symmetry, a point A satisfying the stated condition must lie on the y-axis, say at $(0, k)$, where we may assume $k > 0$. We denote by m and n the slopes of the line segments AM and AN, respectively; so

$$m = \frac{-k}{h - s}, \qquad n = \frac{-k}{h + s},$$

and

$$\tan \angle MAN = \frac{n - m}{1 + mn} = \frac{2sk}{h^2 - s^2 + k^2}.$$

Using (1) to eliminate h, we find

$$\tan \angle MAN = \frac{2sk}{2sr + r^2 + k^2 - d^2}$$

which has a constant value, independent of s, if and only if $k^2 = d^2 - r^2$.

P.G./3 First solution. Note that $1985 = 73 \cdot 26 + 99 - 12$. Now consider the 100 lines:

(i) The 99 lines $x = k$, $k = 1, 2, \ldots, 73$, and $y = k$, $k = 1, 2, \ldots, 26$.

(ii) The one line $y = x + 14$.

The lines in (i) intersect in $73 \cdot 26$ points. The line $y = x + 14$ intersects each of the 99 lines in (i), but twelve of those points, $(1, 15), (2, 16), \ldots, (12, 26)$ are also points of intersection of the 99 lines. Thus the answer is in the affirmative.

Second solution. If 100 lines are drawn in the plane in general position (i.e. with no 2 of them parallel and no 3 of them concurrent), the number of intersection points is $\binom{100}{2} = 4950$. We will show that, by moving some of the lines, we can reduce this number to 1985. We keep the lines non-parallel, but translate some of them so as to form bunches of size $n_1, n_2, \ldots, n_k$ (all $\geqslant 3$), each bunch intersecting in a single point. The lines in a bunch of size n_i, which used to intersect each other in $\binom{n_i}{2}$ points, now intersect in only 1 point. Thus the formation of this bunch has reduced the number of intersections by $\binom{n_i}{2} - 1$. Since we want the total reduction to be $4950 - 1985 = 2965$, we are led to the problem of solving the system

$$n_1 + n_2 + \cdots + n_k \leqslant 100$$

$$\binom{n_1}{2} - 1 + \binom{n_2}{2} - 1 + \cdots + \binom{n_k}{2} - 1 = 2695$$

in integers $n_i \geqslant 3$. An easy calculation shows that the greatest integer of the form $\binom{n}{2} - 1$ less than 2965 is $\binom{77}{2} - 1 = 2925$. Hence we take $n_1 = 77$, and must now make a further reduction of $2965 - 2925 = 40$. Next, $\binom{9}{2} - 1 = 35$, so we take $n_2 = 9$ and must now make a further reduction of $40 - 35 = 5$. Since $\binom{4}{2} - 1 = 5$, we take $n_3 = 4$ to complete the task. (Note that $n_1 + n_2 + n_3 = 77 + 9 + 5 = 91 < 100$, which means there are 9 lines we have not disturbed in this process.)

P.G./4. A line l is called a *supporting line* of F if
  (i) l contains at least one point of F,
  (ii) F is contained in one of the two closed half-planes determined by l.
[Figure 1 shows a closed convex set F and four if its supporting lines.]

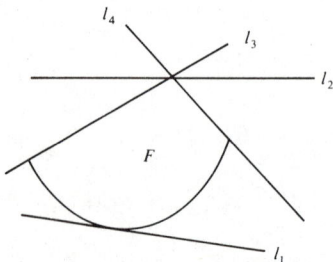

Figure 1

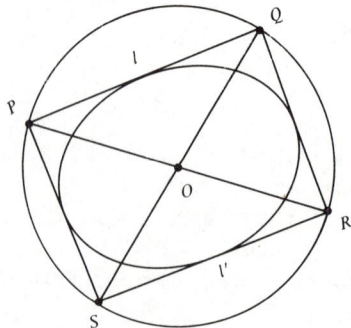

Figure 2

If l is a supporting line of F, and H is a closed half-plane determined by l and containing F, we call H a *supporting half-plane* of F.

We recall the following fundamental facts about closed convex sets F.†

(a) Through each boundary point of F there passes at least one supporting line.

(b) F is the intersection of its supporting half-planes.

Let l be any supporting line of F, and suppose it intersects the given circle in points P and Q. See Figure 2. By assumption, the lines through P and Q perpendicular to l are also supporting lines of F. They intersect the circle in points R and S diametrically opposite P and Q. Again by assumption, the line $l' = RS$ is a supporting line of F. Since l' is the reflection of l in O, we have shown that the set of supporting lines of F is closed under reflection in O. Therefore the set of supporting half-planes of F is also closed under reflection in O, and since F is the intersection of all these half-planes, it too is closed under reflection in O.

Solid Geometry

S.G./ 1 First solution. Let $OABC$ be the tetrahedron, and let $OA = \mathbf{a}$, $OB = \mathbf{b}$, $OC = \mathbf{c}$, and $OH = \mathbf{h}$, where H is the foot of the altitude from O. Then, since $OH \perp \triangle ABC$, $0 = \mathbf{h} \cdot (\mathbf{a} - \mathbf{b}) = \mathbf{h} \cdot (\mathbf{b} - \mathbf{c}) = \mathbf{h} \cdot (\mathbf{c} - \mathbf{a})$, and hence

$$(1) \qquad\qquad \mathbf{h} \cdot \mathbf{a} = \mathbf{h} \cdot \mathbf{b} = \mathbf{h} \cdot \mathbf{c} = m.$$

Since H is the orthocenter of $\triangle ABC$,

$$0 = (\mathbf{h} - \mathbf{a}) \cdot (\mathbf{b} - \mathbf{c}) = (\mathbf{h} - \mathbf{b}) \cdot (\mathbf{c} - \mathbf{a}) = (\mathbf{h} - \mathbf{c}) \cdot (\mathbf{a} - \mathbf{b});$$

so $\mathbf{h} \cdot (\mathbf{b} - \mathbf{c}) = \mathbf{a} \cdot (\mathbf{b} - \mathbf{c})$, and, by (1), $\mathbf{a} \cdot \mathbf{b} = \mathbf{a} \cdot \mathbf{c}$. Similarly, $\mathbf{b} \cdot \mathbf{c} = \mathbf{b} \cdot \mathbf{a}$, thus

$$(2) \qquad\qquad \mathbf{a} \cdot \mathbf{b} = \mathbf{b} \cdot \mathbf{c} = \mathbf{c} \cdot \mathbf{a} = n.$$

The altitude from A to face OBC intersects OH if and only if for some scalar k, we have

$$(k\mathbf{h} - \mathbf{a}) \cdot \mathbf{b} = 0 \quad \text{and} \quad (k\mathbf{h} - \mathbf{a}) \cdot \mathbf{c} = 0.$$

In view of (1) and (2) both are true for $k = n/m$. Thus $(n/m)\mathbf{h}$ clearly lies on the altitude from A. By the same argument, we see that $(n/m)\mathbf{h}$ also lies on the altitudes from B and C.

Second solution. Since OH and AH are $\perp$ to BC, OA is also $\perp$ to BC, and similarly for the two other pairs of opposite sides of $OABC$. A tetrahedron whose opposite edges are orthogonal is said to be *orthocentric*.

†See [52].

It is a known theorem that the four altitudes of an orthocentric tetrahedron are concurrent at a point called (naturally) the *orthocenter*. For a synthetic proof of the theorem, see reference [96, p. 71].

For a more challenging related problem, prove that if one altitude of a tetrahedron intersects two other altitudes, then the four altitudes are concurrent.

S.G./2. Let F_k denote the number of faces of the polyhedron with k edges. Then the sum of all the face angles of the polyhedron is

$$S = \pi(F_3 + 2F_4 + 3F_5 + \cdots).$$

Also

$$2E = 3F_3 + 4F_4 + 5F_5 + \cdots,$$

where E is the total number of edges of the polyhedron. Hence,

$$S/\pi + 2E = 4F_3 + 6F_4 + 8F_5 + \cdots$$

or $S = 2m\pi$ for some integer m. If S' is the sum of the missing face angles, then $0 < S' < 2\pi$ and

$$2m\pi - S' = 5160° = 28\pi + 120°.$$

Thus

$$0 < S' = 2\pi(m - 14 - 1/3) < 2\pi, \qquad m = 15,$$
$$S' = 240° \qquad \text{and} \qquad S = 5160° + 240° = 5400°.$$

S.G./3. Each point of T has six neighbors. So for each allowable set of $(6 + 1)$ points, we want S to contain exactly one of them. This suggests that for S, we consider an appropriate set of points modulo 7. This leads one to define S as the set of points (h, k, l) such that $h + 2k + 3l$ is a multiple of seven. The neighbors of any point (x, y, z) are $(x \pm 1, y, z)$, $(y, y \pm 1, z)$, and $(x, y, z \pm 1)$. Since among the seven consecutive numbers $x + 2y + 3z$, $(x \pm 1) + 2y + 3z$, $x + 2(y \pm 1) + 3z$, and $x + 2y + 3(z \pm 1)$, exactly one of them is divisible by seven, we have obtained a set S with the desired property.

The above result generalizes to lattice points $(x_1, x_2, \ldots, x_n)$ in n-dimensions. Here, an appropriate set S consists of all the points $(h_1, h_2, \ldots, h_n)$ such that $h_1 + 2h_2 + \cdots + nh_n$ is a multiple of $2n + 1$.

S.G./4. We first prove the

LEMMA. *Let A, B, C, D be the vertices of a tetrahedron T, and let M be the center of its circumscribed sphere. T' is the tetrahedron whose vertices are the centroids of the faces of T, and K is the center of the circumsphere S of T'. Then the vector representation of K is*

(1) $$\mathbf{k} = \tfrac{1}{3}(\mathbf{a} + \mathbf{b} + \mathbf{c} + \mathbf{d} - \mathbf{m})$$

(*where* $\mathbf{x}$ *denotes the vector to point X from some common origin O*).

PROOF: The radius vectors from K to the vertices of T' are

$$\tfrac{1}{3}(\mathbf{b} + \mathbf{c} + \mathbf{d}) - \mathbf{k}, \quad \tfrac{1}{3}(\mathbf{c} + \mathbf{d} + \mathbf{a}) - \mathbf{k}, \quad \tfrac{1}{3}(\mathbf{d} + \mathbf{a} + \mathbf{b}) - \mathbf{k},$$

$$\tfrac{1}{3}(\mathbf{a} + \mathbf{b} + \mathbf{c}) - \mathbf{k}.$$

If $\mathbf{k}$ is given by (1), we express them as

$$\tfrac{1}{3}(\mathbf{m} - \mathbf{a}), \quad \tfrac{1}{3}(\mathbf{m} - \mathbf{b}), \quad \tfrac{1}{3}(\mathbf{m} - \mathbf{c}), \quad \tfrac{1}{3}(\mathbf{m} - \mathbf{d}),$$

and note that these are parallel to, and one-third as long as the radius vectors from M to the vertices of T.

In the given problem, set $M = O$, so $|\mathbf{m}| = 0$ and $|\mathbf{a}| = |\mathbf{b}| = |\mathbf{c}| = |\mathbf{d}| = 1$, so the radius of S has length $\tfrac{1}{3}$.

The distance $|\mathbf{k}|$ between the centers of the spheres, by (1), satisfies

$$9|\mathbf{k}|^2 = |\mathbf{a} + \mathbf{b} + \mathbf{c} + \mathbf{d}|^2$$

$$= |\mathbf{a}|^2 + |\mathbf{b}|^2 + |\mathbf{c}|^2 + |\mathbf{d}|^2$$

$$+ 2(\mathbf{a} \cdot \mathbf{b} + \mathbf{a} \cdot \mathbf{c} + \mathbf{a} \cdot \mathbf{d} + \mathbf{b} \cdot \mathbf{c} + \mathbf{b} \cdot \mathbf{d} + \mathbf{c} \cdot \mathbf{d}).$$

We use the identity

$$2\mathbf{u} \cdot \mathbf{v} = |\mathbf{u}|^2 + |\mathbf{v}|^2 - |\mathbf{u} - \mathbf{v}|^2$$

and the fact that $\mathbf{a}, \mathbf{b}, \mathbf{c}, \mathbf{d}$ are unit vectors to write

$$(2) \qquad 9|\mathbf{k}|^2 = 16 - \text{sum of the squares of the edges of } T.$$

Geometric Inequalities

G.I./ 1. We denote the vectors from O to P_i and from O to Q by $\mathbf{p}_i$ and $\mathbf{q}$, respectively. We shall consider the somewhat more general situation where P_i are points on a unit sphere of any dimension with center O, and where a weight w_i is located at each point P_i so that the centroid of these weights is at O. We shall show that then

$$\sum_{i=1}^{n} w_i \overline{P_i Q} \geq \sum_{i=1}^{n} w_i$$

for every point Q.

In vector notation, our hypothesis is

$$|\mathbf{p}_i| = 1, \ i = 1, 2, \ldots, n \quad \text{and} \quad \sum_{i=1}^{n} w_i \mathbf{p}_i = \mathbf{o},$$

where $\mathbf{o}$ is the zero vector, and the desired conclusion is

$$\sum_{i=1}^{n} w_i |\mathbf{q} - \mathbf{p}_i| \geq \sum_{i=1}^{n} w_i \qquad \text{for all } \mathbf{q}.$$

The definition of dot product, $\mathbf{u} \cdot \mathbf{v} = |\mathbf{u}| \, |\mathbf{v}| \cos(\mathbf{u}, \mathbf{v})$, yields the inequality $|\mathbf{u}| \, |\mathbf{v}| \geq \mathbf{u} \cdot \mathbf{v}$, where $=$ holds if and only if $\mathbf{u}, \mathbf{v}$ have the same direction or one of them is the zero vector. We apply this as follows: Since $|\mathbf{p}_i| = 1$,

(1) $\quad |\mathbf{q} - \mathbf{p}_i| = |\mathbf{p}_i| \, |\mathbf{p}_i - \mathbf{q}| \geqslant \mathbf{p}_i \cdot (\mathbf{p}_i - \mathbf{q}) = 1 - \mathbf{p}_i \cdot \mathbf{q}, \qquad i = 1, 2, \ldots, n.$

Multiplying this by w_i and then adding all n of these relations, we obtain

$$\sum_{i=1}^{n} w_i |\mathbf{q} - \mathbf{p}_i| \geqslant \sum_{i=1}^{n} w_i (1 - \mathbf{q} \cdot \mathbf{p}_i) = \sum_{i=1}^{n} w_i - \mathbf{q} \cdot \sum_{i=1}^{n} w_i \mathbf{p}_i = \sum_{i=1}^{n} w_i.$$

Equality holds if and only if it holds in each of the relations (1), i.e. if and only if $\mathbf{q}$ lies on each $\mathbf{p}_i$. This can be only if $\mathbf{q} = \mathbf{o}$, that is, only if Q is the center O of the unit sphere.

For proofs of other geometric inequalities via vectors, see Amer. Math. Monthly, 77 (1970), pp. 1051–1065.

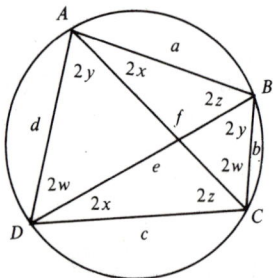
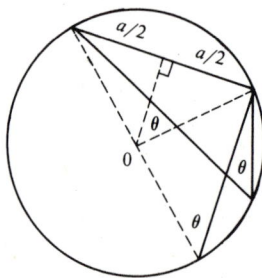

Figure 1   Figure 2

G.I./2. In Figure 1 we have labelled the relevant lengths and angles. By the triangle inequality, we have

$$a + d > e, \quad d + c > f, \quad c + b > e, \quad b + a > f.$$

Thus, $a + b + c + d > e + f$, or $u > 0$.

At least one of the angles of quadrilateral $ABCD$ is $\leqslant \pi/2$, say the angle at A; so we may assume $x + y \leqslant \pi/4$. Also, $x + y + z + w = \pi/2$. We may also assume, without loss of generality, that $z \leqslant w$.

Since the radius of the circle is 1, we have (see Figure 2)

$$a = 2 \sin 2w, \quad b = 2 \sin 2x, \quad c = 2 \sin 2y, \quad d = 2 \sin 2z,$$
$$e = 2 \sin 2(x + y), \quad f = 2 \sin 2(w + x);$$

so the inequality $2 > u$ is equivalent to

(1) $\quad 1 + \sin 2(x + y) + \sin 2(w + x) > \sin 2x + \sin 2y + \sin 2z + \sin 2w.$

We shall prove the sharper inequality

(2)
$$\sin 2(x + y) + \sin 2(y + z) + \sin 2(z + x)$$
$$> \sin 2x + \sin 2y + \sin 2z + \sin 2(x + y + z)$$

in which 1 in (1) has been replaced by $\sin 2(z + x)$, (and $\sin 2(w + x)$, $\sin 2w$, by the equivalent quantities $\sin 2(y + z)$, $\sin 2(x + y + z)$, respectively).

Using the identity $\sin 2\alpha + \sin 2\beta = 2\sin(\alpha + \beta)\cos(\alpha - \beta)$ and the addition formula for $\sin$ in (2), we get the equivalent inequality

$$\sin(x + y)\cos(x + y) + \sin(x + y + 2z)\cos(x - y)$$
$$> \sin(x + y)\cos(x - y) + \sin(x + y + 2z)\cos(x + y)$$

which simplifies to

$$[\sin(x + y + 2z) - \sin(x + y)][\cos(x - y) - \cos(x + y)] > 0.$$

This is valid because $x + y < x + y + 2z \leqslant \pi/2$.

If we allow degenerate quadrilaterals, the inequality $u < 2$ and the sharper inequality (2) can become equalities; take, for example, $ABCD$ to be a right triangle with C and D coalescing at the vertex of the right angle.

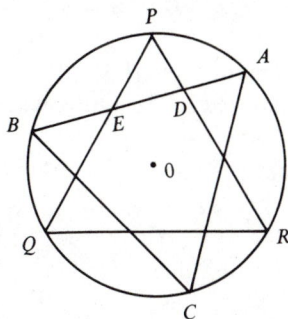

G.I./3. Denote the triangles by ABC and PQR, and let D and E be the points of intersection of AB with PR and AB with PQ, respectively. Then by rotational symmetry, the entire figure is symmetric about the line OD, and also the line OE. Moreover

$$K = \text{Area } ABC - 3 \text{ Area } PDE.$$

So K will be a minimum when $\triangle PDE$ has maximum area. Intuitively, one expects this when P is the midpoint of arc AB. To prove this, note that $PD = AD$, $PE = BE$, so that $\triangle PDE$ has the constant perimeter $AB = r\sqrt{3}$. It is an elementary result† that of all triangles with given perimeter, the equilateral triangle has the maximum area. Consequently, $\triangle PDE$ has maximum area when P is the midpoint of arc AB. In this case the sides of

---

†For a proof, see e.g. Chapter 2 of *Geometric Inequalities* by N. Kazarinoff, NML vol. 4, in particular Theorem 11A.

$\triangle PDE$ are $\frac{1}{3}$ as long as the sides of $\triangle ABC$, so the area of $\triangle PDE$ is $\frac{1}{9}$ that of $\triangle ABC$.

$$K \geqslant (\text{Area } ABC)\left(1 - \frac{3}{9}\right) = \frac{2}{3}\left(r\sqrt{3}\right)^2 \frac{\sqrt{3}}{4} = r^2\sqrt{3}/2.$$

In a similar fashion, one can obtain the analogous area inequality for two regular n-gons inscribed in a circle. K will be a minimum when one of the n-gons can be gotten from the other one by a rotation of π/n about the center.

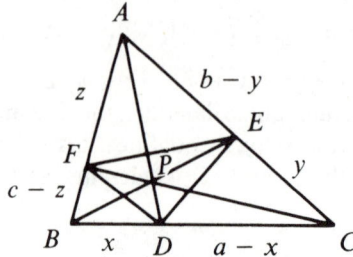

G.I./4 First solution. Denoting by $[UVW]$ the area of a $\triangle UVW$, and referring to the lengths as labelled in the figure, we see that

$$(1) \qquad [DEF] = [ABC] - [BDF] - [DCE] - [EAF].$$

Since $[ABC] = \frac{1}{2}ac \sin B$, we have $\frac{1}{2}\sin B = [ABC]/ac$. Hence $[BDF] = \frac{1}{2}x(c - z)\sin B = [ABC]x(c - z)/ac$, and analogous expressions give the other areas. Setting these into (1), we get

$$(2) \qquad \begin{aligned} [DEF] &= [ABC]\left(1 - \frac{x(c-z)}{ac} - \frac{y(a-x)}{ba} - \frac{z(b-y)}{cb}\right) \\ &= [ABC](1 - u(1-w) - v(1-u) - w(1-v)), \end{aligned}$$

where $u = x/a$, $v = y/b$, $w = z/c$. We wish to maximize the second factor on the right of (2) which we may write in the form

$$(3) \qquad F = (1-u)(1-v)(1-w) + uvw.$$

By Ceva's theorem† u, v, w satisfy

$$(4) \qquad uvw = (1-u)(1-v)(1-w).$$

Denote uvw by G; then by (3), $F = 2G$, and F will reach its maximum when G does, or when

$$(5) \qquad G^2 = u(1-u)v(1-v)w(1-w)$$

does. The maximum of G^2 is easy to see by inspection, either from the fact

---

†See Glossary; for a proof, see e.g. *Geometry Revisited* by H. S. M. Coxeter and S. L. Greitzer, NML vol. 19, p. 4.

that $s(1 - s) \leqslant \frac{1}{4}$, or from the AM − GM inequality which yields $G^{1/3} \leqslant \frac{1}{2}$ and occurs when $u = v = w = \frac{1}{2}$ and satisfies (4). Thus

$$\max[DEF] = \tfrac{1}{4}[ABC]$$

and is reached if and only if P is the centroid of $\triangle ABC$.

Generalization and Second Solution. We now state and prove a generalization† which, in the case $n = 3$, yields a second solution to our problem.

THEOREM. *Let* $V_0, V_1, \ldots, V_n$ *denote the* $n + 1$ *vertices of an n-dimensional simplex S in E^n, and $V_0', V_1', \ldots, V_n'$ the $n + 1$ vertices of a simplex S' inscribed in S so that the segments V_iV_i' $(i = 0, 1, \ldots, n)$ all meet at a point P within S. Then*

$$\mathrm{Vol}\, S \geqslant n^n \mathrm{Vol}\, S',$$

with equality if and only if P is the centroid of S.

PROOF: Let $\mathbf{v}_i$ denote the $n + 1$ vectors from some origin to the vertices V_i, $i = 0, 1, \ldots, n$. The point P in the simplex S can be represented as a *convex combination* of the $\mathbf{v}_i$, that is, as a vector of the form

$$(1) \qquad \mathbf{p} = \lambda_0\mathbf{v}_0 + \lambda_1\mathbf{v}_1 + \cdots + \lambda_n\mathbf{v}_n, \qquad \lambda_i \geqslant 0, \ \sum_{i=0}^{n} \lambda_i = 1.$$

The λ_i are called *barycentric coordinates* of P in S, and it can be shown that $\mathrm{Vol}.(V_0, \ldots, V_{k-1}, P, V_{k+1}, \ldots, V_n)/\mathrm{Vol}.(V_0, V_1, \ldots, V_n) = \lambda_k$.

The vertices V_i' of S' are then represented by the vectors

$$(2) \qquad \mathbf{v}_i' = \frac{1}{1 - \lambda_i}\mathbf{p} - \frac{\lambda_i}{1 - \lambda_i}\mathbf{v}_i, \qquad i = 0, 1, \ldots, n.$$

We now choose V_0 to be the origin, so $\mathbf{v}_0 = \mathbf{o}$.

The signed volumes of S and S' are given by

$$\mathrm{Vol}\, S = \frac{1}{n!}[\mathbf{v}_1, \mathbf{v}_2, \ldots, \mathbf{v}_n], \qquad \mathrm{Vol}\, S' = \frac{1}{n!}[\mathbf{v}_1' - \mathbf{v}_0', \mathbf{v}_2' - \mathbf{v}_0', \ldots, \mathbf{v}_n' - \mathbf{v}_0'],$$

where $[\mathbf{u}_1, \mathbf{u}_2, \ldots, \mathbf{u}_n]$ denotes the determinant with columns $\mathbf{u}_i$. By means of relation (2), the latter can be written

$$\mathrm{Vol}.\, S' = \frac{1}{n!}[r_1\mathbf{v}_1 + s_1\mathbf{p}, r_2\mathbf{v}_2 + s_2\mathbf{p}, \ldots, r_n\mathbf{v}_n + s_n\mathbf{p}],$$

where

$$r_i = \frac{-\lambda_i}{1 - \lambda_i}, \qquad s_i = \frac{\lambda_i - \lambda_0}{(1 - \lambda_i)(1 - \lambda_0)}.$$

†M. S. Klamkin, *A Volume Inequality*, Univ. Beo. Publ. Elek. Fac. Ser. Mat. i. Fiz., no. 357–380 (1971) pp.3–4.

We note that a determinant is a linear function of each of its columns. Starting with the first column, we get

$$[r_1 v_1 + s_1 p, r_2 v_2 + s_2 p, \ldots, r_n v_n + s_n p]$$

$$= r_1 [v_1, r_2 v_2 + s_2 p, \ldots, r_n v_n + s_n p] + s_1 [p, r_2 v_2 + s_2 p, \ldots, r_n v_n + s_n p].$$

Similarly we break up these determinants with respect to their second columns, the result with respect to their third columns, etc. Since the determinant of a matrix with two identical columns is 0, we get $n + 1$ terms

$$r_1 r_2 \cdots r_n [v_1, v_2, \ldots, v_n] + \sum_{i=1}^{n} s_i \prod_{k \neq i} r_k [v_1, \ldots, v_{i-1}, p, \ldots v_n].$$

Substituting from (1) for p, we finally arrive at the following expression for $|\text{Vol. } S'|$ in terms of $|\text{Vol. } S|$:

$$
(3) \qquad |\text{Vol } S'| = |\text{Vol } S| \left\{ \prod_{i=1}^{n} r_i \right\} \left(1 + \frac{s_1 \lambda_1}{r_1} + \frac{s_2 \lambda_2}{r_2} + \cdots + \frac{s_n \lambda_n}{r_n} \right)
$$

$$
= \frac{n \lambda_0 \lambda_1 \cdots \lambda_n}{(1 - \lambda_0)(1 - \lambda_1) \cdots (1 - \lambda_n)} |\text{Vol.} S|.
$$

The maximum of the coefficient of $|\text{Vol.} S|$ in the last expression subject to $\sum_{i=0}^{n} \lambda_i = 1$ is easily found with the help of the A.M.-G.M. inequality

$$
\frac{1 - \lambda_i}{n} = \frac{\lambda_0 + \lambda_1 + \cdots + \lambda_n - \lambda_i}{n} \geq \left(\frac{\lambda_0 \lambda_1 \cdots \lambda_n}{\lambda_i} \right)^{1/n}
$$

for $i = 0, 1, \ldots, n$. The product of these $n + 1$ inequalities yields

$$
\prod_{i=0}^{n} (1 - \lambda_i) \geq n^{n+1} \prod_{i=0}^{n} \lambda_i
$$

with equality if and only if $\lambda_i = 1/(n + 1)$, $i = 0, 1, \ldots, n$. Combining this with (3) gives the inequality stated in the theorem with equality if and only if all the barycentric coordinates of P are equal, i.e. if and only if P is the centroid of S.

G.I./5. Let a, b, c be vectors corresponding to three coterminal edges of the parallelepiped. We have to prove that

(1) $|a + b + c| + |b + c - a| + |c + a - b| + |a + b - c| \geq 2(|a| + |b| + |c|)$.

Set

$$u = b + c - a, \qquad v = c + a - b, \qquad w = a + b - c;$$

then (1) takes the form

(2) $|u + v + w| + |u| + |v| + |w| \geq |v + w| + |w + u| + |u + v|$.

Inequality (2) is due to E. Hlawka, [108, p. 172] and based on the identity

$$\big(|\mathbf{u}| + |\mathbf{v}| + |\mathbf{w}| - |\mathbf{v} + \mathbf{w}| - |\mathbf{w} + \mathbf{u}| - |\mathbf{u} + \mathbf{v}| + |\mathbf{u} + \mathbf{v} + \mathbf{w}|\big)$$
$$\times\big(|\mathbf{u}| + |\mathbf{v}| + |\mathbf{w}| + |\mathbf{u} + \mathbf{v} + \mathbf{w}|\big)$$
$$= \big(|\mathbf{v}| + |\mathbf{w}| - |\mathbf{v} + \mathbf{w}|\big)\big(|\mathbf{u}| - |\mathbf{v} + \mathbf{w}| + |\mathbf{u} + \mathbf{v} + \mathbf{w}|\big)$$

(3)
$$+ \big(|\mathbf{w}| + |\mathbf{u}| - |\mathbf{w} + \mathbf{u}|\big)\big(|\mathbf{v}| - |\mathbf{w} + \mathbf{u}| + |\mathbf{u} + \mathbf{v} + \mathbf{w}|\big)$$
$$+ \big(|\mathbf{u}| + |\mathbf{v}| - |\mathbf{u} + \mathbf{v}|\big)\big(|\mathbf{w}| - |\mathbf{u} + \mathbf{v}| + |\mathbf{u} + \mathbf{v} + \mathbf{w}|\big).$$

For, by the triangle inequality, each factor on the right of (3) is ≥ 0, so the left side of (3) is non-negative, which implies (2).

A second proof is based on the following theorem of F. W. Levi [108, p. 175]:

Let a_{ij} $(i = 1,\ldots, p;\ j = 1,\ldots, r)$, b_{ij} $(i = 1,\ldots, q;\ j = 1,\ldots, r)$ denote given real numbers. If the inequality

(4)
$$\sum_{i=1}^{p}\left|\sum_{j=1}^{r} a_{ij}v_j\right| \leqq \sum_{i=1}^{q}\left|\sum_{j=1}^{r} b_{ij}v_j\right|$$

is true for all sets of real numbers $v_1,\ldots, v_r$, then it is also true for n-component vectors $\mathbf{v}_1,\ldots,\mathbf{v}_r$.

We give the proof for $n = 2$ dimensions. Let $\mathbf{u}(\theta)$ denote the unit vector forming an angle θ with the positive x-axis. Then for any vector $\mathbf{v}$,

(5)
$$\int_0^{2\pi}|\mathbf{v} \cdot \mathbf{u}(\theta)|\, d\theta = 4|\mathbf{v}|.$$

By assumption we get a true inequality if in (4) we replace $\mathbf{v}_1,\ldots,\mathbf{v}_r$ by their components in the direction $\mathbf{u}(\theta)$:

$$\sum_{i=1}^{p}\left|\sum_{j=1}^{r} a_{ij}\mathbf{v}_j \cdot \mathbf{u}(\theta)\right| \leq \sum_{i=1}^{q}\left|\sum_{j=1}^{r} b_{ij}\mathbf{v}_j \cdot \mathbf{u}(\theta)\right|.$$

Integrating with respect to θ and applying (5), we get (4).

The proof in more dimensions is the same except that (5) becomes

(5)′
$$\int_S |\mathbf{v} \cdot \mathbf{u}(S)|\, dS = k|\mathbf{v}|,$$

where the integral is over the surface of an n-dimensional sphere, and k is a constant independent of $\mathbf{v}$. This follows since we can write $\mathbf{v} = |\mathbf{v}|\mathbf{v}'$, where $\mathbf{v}'$ is a unit vector in the direction of $\mathbf{v}$. Then

$$\int_S |\mathbf{v} \cdot \mathbf{u}(S)|\, dS = |\mathbf{v}|\int_S |\mathbf{v}' \cdot \mathbf{u}(S)|\, dS,$$

and the last integral is independent of $\mathbf{v}'$ by spherical symmetry.

To prove our original inequality it suffices, by Levi's theorem for $n = 3$ dimensions, to prove

$$
(6) \quad
\begin{aligned}
|a + b + c| + |a + b - c| + |a - b + c| + |-a + b + c| \\
\geq 2|a| + 2|b| + 2|c|
\end{aligned}
$$

for real numbers a, b, c. First observe that changing the sign of any of a, b, c just permutes the terms on the left side and leaves the right side unchanged, so we may assume without loss of generality that a, b and c are non-negative.

Now

$$
(6)' \quad
\begin{aligned}
(a + b + c) + (a + b - c) + (a - b + c) + (b + c - a) \\
= 2a + 2b + 2c.
\end{aligned}
$$

Since $a, b, c \geq 0$, the right sides of (6) and (6)' are equal, while the left side of (6) is $\geq$ the left side of (6)'. This completes the proof of (6).

F. W. Levi's theorem yields the following extension of inequality (1) for r vectors in a space of any number of dimensions

$$
(7) \quad \sum |\pm \mathbf{a}_1 \pm \mathbf{a}_2 \pm \cdots \pm \mathbf{a}_r| \geq 2 \binom{r-1}{t} \sum_{i=1}^{r} |\mathbf{a}_i|,
$$

where $t = \lfloor \frac{1}{2}(r - 1) \rfloor$, and the sum on the left hand side is taken over all combinations of $+$ and $-$. For other extensions and related inequalities, see [108, pp. 171–176] and Amer. Math. Monthly 82 (1975) pp. 829–830.

The first step in the proof is again a reduction to the 1-dimensional case, and to the case $a_i \geq 0$, $i = 1, \ldots, r$. Next we construct an identity similar to (6)', as follows: Let $s_i = \pm 1$ ($i = 1, \ldots, r$) and

$$
s = \begin{cases}
1 & \text{if } \Sigma s_i > 0 \\
0 & \text{if } \Sigma s_i = 0 \\
-1 & \text{if } \Sigma s_i < 0.
\end{cases}
$$

Then we claim that

$$
(8) \quad \sum_{i=1}^{r} s(s_1 a_1 + \cdots + s_r a_r) = 2 \binom{r-1}{t} \sum_{i=1}^{r} a_i,
$$

where the sum is over all combinations of $\pm$ signs. The factor s causes the majority of signs in each nonzero term to be positive. The symmetry of the left hand side implies that it is equal to some constant times the sum on the right. We have to show the constant is as shown.

The coefficient of a_1 in the sum on the left side of (8) is equal to the number of terms in which s_1 is equal to the majority of the s_i (including s_1), minus the number of terms in which s_1 is opposite to the majority of the s_i.

Suppose r is odd. To each fixed value of $s_2, \ldots, s_r$ we have two possible values of s_1. If the number of positive values among the $s_2, \ldots, s_r$ is not equal to the number of negative values, then they differ by at least 2, since $r - 1$ is even. Hence one of the two possible values of s_1 will be in the majority and one in the minority, and the contributions of the two terms to the coefficient of a_1 cancel. If however

$\sum_2^r s_i = 0$, then s_1 has the deciding vote, so to speak, and both terms make a total contribution of 2. The number of ways in which exactly $\frac{1}{2}(r-1)$ of $s_1, \ldots, s_r$ can be chosen positive is $\binom{r-1}{t}$. This proves (8) when r is odd.

The case when r is even is left as an exercise. The proof is completed by taking absolute values on the left side of (8).

The constant we obtained is best possible. It is attained when for instance, $a_1 = a_2 = \cdots = a_r = 1$.

Remark. Our problem concerning the lengths of diagonals is related to the following voting problem: I and $r-1$ others vote for one of two candidates, and everyone makes his decision by flipping a fair coin. What is the probability that my candidate wins a majority of votes?

When r is odd, there is always a winner, and we found that the favorable possibilities exceed the unfavorable cases by $2\binom{r-1}{t}$. Since the sum of all possibilities is 2^r, we can find the favorable cases from these two equations.

When r is even, I can be on the winning side only if a majority of the others voted for my candidate. Since an odd number of people cannot be deadlocked, the probability of being on the winning side is $\frac{1}{2}$ in this case. However, the probability of being on the losing side is $< \frac{1}{2}$, because I can cause a deadlock if a candidate has a majority of only 1 among the others.

G.I./6. Partition the cube into 64 cubes of edge-length $1/4$. Since $1985 = 31 \cdot 64 + 1$, there is at least one small cube containing at least 32 points. In this cube, the longest possible segment is a diagonal which has length $\sqrt{3}/4$. Since there are at most four segments whose length is exactly $\sqrt{3}/4$, the perimeter of the polygon with these 32 points as vertices is $< 32\sqrt{3}/4 = 8\sqrt{3}$.

G.I./7 First solution. Let the center of the sphere be the origin. Let the vectors from the center to the vertices be $v_1, \ldots, v_4$. We shall derive the inequality from a formula that expresses the v_i, which we know have length 1, in terms of the edges of the tetrahedron.

The condition that the center of the sphere is inside the tetrahedron is equivalent to the existence of non-negative numbers $c_1, \ldots, c_4$ so that

$$(1) \qquad c_1v_1 + \cdots + c_4v_4 = 0 \quad \text{and} \quad c_1 + \cdots + c_4 = 1.$$

These equations make it possible to express the individual vectors v_i in terms of the edges $v_j - v_k$. We have

$$(2) \quad v_1 = c_2(v_1 - v_2) + c_3(v_1 - v_3) + c_4(v_1 - v_4) + (c_1v_1 + \cdots + c_4v_4)$$
$$= c_2(v_1 - v_2) + c_3(v_1 - v_3) + c_4(v_1 - v_4).$$

Let $|v_i - v_j| = x_{ij}$. Taking absolute values in (2) gives

$$c_2x_{12} + c_3x_{13} + c_4x_{14} \geq 1,$$

and we have three similar inequalities obtained by cyclic permutation of the indices. The sum of the four inequalities is

$$(3) \qquad \sum_{i<j}(c_i + c_j)x_{ij} \geq 4$$

or

$$(4) \qquad S = \sum_{i=1}^{4} c_i \sum_{j \neq i} x_{ij} = \sum_{i=1}^{4} c_i X_i \geq 4,$$

where X_i denotes the sum of the edges incident to the i-th vertex.

Since all the v_i have length 1, (1) implies that none of the coefficients c_i is greater than the sum of the other three, otherwise the sum could not be the zero vector. This implies

$$(5) \qquad c_i \leq \tfrac{1}{2}, \qquad\qquad i = 1,2,3,4.$$

We may assume without loss of generality that the vertices are numbered so that

$$(6) \qquad X_1 \geq X_2 \geq X_3 \geq X_4.$$

We claim that

$$(7) \qquad S' = \tfrac{1}{2}X_1 + \tfrac{1}{2}X_2 \geq c_1 X_1 + c_2 X_2 + c_3 X_3 + c_4 X_4 = S,$$

because S' is obtained from S by concentrating as much of the weight factors as we can on the largest X. We can verify (7) algebraically as follows:

$$S' - S = \left(\tfrac{1}{2} - c_1\right)X_1 + \left(\tfrac{1}{2} - c_2\right)X_2 - c_3 X_3 - c_4 X_4.$$

From the second equation in (1)

$$(8) \qquad \left(\tfrac{1}{2} - c_1\right)X_2 + \left(\tfrac{1}{2} - c_2\right)X_2 - c_3 X_2 - c_4 X_2 = 0.$$

Subtracting this from the previous equation we get

$$S' - S = \left(\tfrac{1}{2} - c_1\right)(X_1 - X_2) + c_3(X_2 - X_3) + c_4(X_2 - X_4),$$

and by (5) and (6) all the terms in this sum are non-negative.

From (4) and (7) we get

$$(9) \qquad X_1 + X_2 \geq 8.$$

On the left we have the sum of all the edges except x_{34}, and the edge x_{12} occurs twice. Since the tetrahedron is in a unit sphere, $x_{12} \leq 2$. Hence the sum of the 5 edges other than x_{34} is at least 6, as claimed. Observe that we have proved a little more than we had set out to prove. We have shown that the sum of the *five longest* edges of the tetrahedron is at least 6.

This solution can clearly be generalized; but we give a somewhat different solution in the n-dimensional case and mention some further extensions.

Second solution. We give a generalization† by considering r unit vectors in Euclidean n-space E^n.

THEOREM: *If $v_1, v_2, \ldots, v_r$ are unit vectors in E^n and the origin is contained in their convex hull, then*

$$\text{(1)} \qquad \sum |v_i - v_j| \geqslant 2(r - 1),$$

where the sum is over $1 \leqslant i < j \leqslant r$.

COROLLARY. *The total edge length of a simplex inscribed in a unit sphere in E^n and containing its center is $\geqslant 2n$.*

PROOF: The following identity holds for arbitrary vectors:

$$\sum_{1 \leqslant i < j \leqslant r} |v_i - v_j|^2 = \frac{1}{2} \sum_{i=1}^{r} \sum_{j=1}^{r} |v_i - v_j|^2$$

$$= r \sum_{i=1}^{r} |v_i|^2 - \left(\sum_{i=1}^{r} v_i \right) \cdot \left(\sum_{j=1}^{r} v_j \right).$$

For unit vectors, we get

$$\text{(2)} \qquad \sum_{i<j} |v_i - v_j|^2 = r^2 - \left| \sum_{i=1}^{r} v_i \right|^2.$$

By hypothesis, the convex hull of the v_i contains the origin; this means that the zero vector is a convex combination of the v_i,‡ that is

$$\text{(3)} \qquad \sum_{i=1}^{r} a_i v_i = \mathbf{0}$$

for some a_i satisfying

$$\text{(3')} \qquad a_i \geqslant 0, \quad i = 1, 2, \ldots, r, \quad \text{and} \quad \sum_{i=1}^{r} a_i = 1.‡$$

We can solve (3) for any j and obtain

$$-a_j v_j = \sum_{i \neq j} a_i v_i;$$

Taking absolute values of both sides, applying the triangle inequality and using (3'), we obtain

$$a_j = \left| \sum_{i \neq j} a_i v_i \right| \leqslant \sum_{i \neq j} a_i = 1 - a_j$$

---

†See G. D. Chakerian, M. S. Klamkin, *Inequalities for sums of distances*, Amer. Math. Monthly, 80 (1973), pp. 1009–1017.

‡See C. Davis, *Theory of positive linear dependence*, Amer. J. Math, 76 (1954), pp. 733–746.

or

$$a_j \leqslant \tfrac{1}{2}.$$

Since $\Sigma a_i \mathbf{v}_i = \mathbf{0}$ by (3), we may write

$$\sum_{i=1}^{r} \mathbf{v}_i = \sum_{i=1}^{r} (1 - 2a_i)\mathbf{v}_i.$$

Then we have, since the coefficients on the right are non-negative,

$$(4) \qquad \left| \sum_{i=1}^{r} \mathbf{v}_i \right| \leqslant \sum_{i=1}^{r} |(1 - 2a_i)\mathbf{v}_i| = \sum_{i=1}^{r} (1 - 2a_i) = r - 2.\dagger$$

Setting (4) into (2) gives

$$(5) \qquad \sum_{i<j} |\mathbf{v}_i - \mathbf{v}_j|^2 \geqslant r^2 - (r - 2)^2 = 4r - 4.$$

Since each $|\mathbf{v}_i| = 1$, the triangle inequality gives $|\mathbf{v}_i - \mathbf{v}_j| \leqslant |\mathbf{v}_i| + |\mathbf{v}_j| = 2$, and this implies

$$\sum_{i<j} |\mathbf{v}_i - \mathbf{v}_j|^2 \leqslant 2 \sum_{i<j} |\mathbf{v}_i - \mathbf{v}_j|.$$

This together with (5) gives the desired inequality

$$\sum |\mathbf{v}_i - \mathbf{v}_j| \geqslant 2(r - 1).$$

Equality holds if and only if $r - 1$ of the vectors are equal and the remaining one their negative.

Finally, here is another related result:‡ Let K be a plane convex curve containing the origin in its interior, and let $\mathbf{v}_1, \mathbf{v}_2, \ldots, \mathbf{v}_r$ be points on K whose convex hull contains the origin. Then if s is the length of the minimum chord of K containing the origin,

$$\sum_{i<j} |\mathbf{v}_i - \mathbf{v}_j| \geqslant s(r - 1).$$

Inequalities

I/1. It is not hard to see that a geometric equivalent of the problem is to maximize the sum of the squares of the sides of a polygon inscribed in a given circle. Here the angles subtended by the sides of the polygon at the center of the circle are $2\theta_i$. We will show that the maximum is achieved by an inscribed equilateral triangle.

 If n (the number of vertices of the polygon) is > 3, then at least one angle of the polygon is $\geqslant 90°$ (since the average angle $= (n - 2)\pi/n$).

---

†See M. S. Klamkin and D. J. Newman, An inequality for the sum of unit vectors, Univ. Beo. Publ. Elek. Fac., Ser. Mat. i. Fiz., no. 338–52 (1971) pp. 47–48.
 ‡See footnote† on page 93.

Then if A, B, C are consecutive vertices of the polygon with $\angle B \geqslant 90°$, it follows by the Law of Cosines that $AB^2 + BC^2 = AC^2 + 2AB \cdot BC \cos B$. Since $\cos B \leqslant 0$, $AB^2 + BC^2 \leqslant AC^2$. Consequently, S will be increased or left the same if we let B coincide with A or C which is equivalent to setting a $\theta_i = 0$. Continuing this process, we see that S will be greatest for either $n = 3$ or $n = 2$. For $n = 2$, S_{max} is clearly 2.

For $n = 3$, we have the problem of maximizing $S_3 = \sin^2\theta_1 + \sin^2\theta_2 + \sin^2\theta_3$, where the θ_i are angles of a triangle. Now using the identities

$$2\sin^2\theta = 1 - \cos 2\theta$$

and

$$\cos 2\theta_1 + \cos 2\theta_2 + \cos 2\theta_3 = -1 - 4\cos\theta_1\cos\theta_2\cos\theta_3$$

(note that $\Sigma \cos 2\theta_i = 2\cos(\theta_1 + \theta_2)\cos(\theta_1 - \theta_2) + \cos 2(\theta_1 + \theta_2)$

$$= 2\cos(\theta_1 + \theta_2)[\cos(\theta_1 - \theta_2) + \cos(\theta_1 + \theta_2)] - 1$$

$$= -4\cos\theta_3\cos\theta_1\cos\theta_2 - 1,$$

see also Glossary, *Trigonometric Identities*), we have

$$S_3 = 2 + 2\cos\theta_1\cos\theta_2\cos\theta_3.$$

Since $\theta_1 + \theta_2 + \theta_3 = \pi$ and $\theta_i \geqslant 0$, at most one angle is $\geqslant \pi/2$. Since in this case $S_3 \leqslant 2$, it is clear that for a maximum all angles must be acute.

We shall use the

LEMMA. *In the interval* $[0, \pi/2),$ $\ln\cos\theta$ *is concave.*

To prove the lemma, we can either use the result: A function is concave in an interval if its second derivative is $\leqslant 0$ there (which $\ln\cos x$ is:

$$d^2(\ln\cos x)/dx^2 = -1/\cos^2 x < 0;)$$

or we can avoid calculus as follows: To say that $\ln\cos x$ is concave in $[0, \pi/2)$ means

$$\frac{1}{2}(\ln\cos x + \ln\cos y) \leqslant \ln\cos\left(\frac{x+y}{2}\right) \quad \text{for } x, y \in [0, \pi/2).$$

But this is equivalent to

$$\cos x \cos y \leqslant \cos^2\frac{x+y}{2} = \frac{1 + \cos(x+y)}{2}$$

or

$$2\cos x \cos y - \cos(x+y) = \cos(x-y) \leqslant 1,$$

which is true.

We now apply Jensen's inequality† and obtain

$$\frac{1}{3}(\ln\cos\theta_1 + \ln\cos\theta_2 + \ln\cos\theta_3) \leqslant \ln\cos\frac{\theta_1 + \theta_2 + \theta_3}{3},$$

---

†See Hungarian Problem Book II, NML vol. 12, MAA (1963), p. 73 ff. or the Glossary.

or

$$\cos\theta_1 \cos\theta_2 \cos\theta_3 \leqslant \left(\cos\frac{\pi}{3}\right)^3 = \frac{1}{8},$$

with equality if and only if $\theta_1 = \theta_2 = \theta_3 = 60°$.

Remark. For several other proofs for the max of S_3, see [106]. More generally, it is also known that for angles of a triangle,

$$\left[(\sin^k\theta_1 + \sin^k\theta_2 + \sin^k\theta_3)/3\right]^{1/k} \leqslant \sqrt{3}/2,$$

if and only if

$$k \leqslant (\log 9 - \log 4)/(\log 4 - \log 3) \approx 2.81884$$

(see A. Makowski, J. Berkes, Elem. der Math. 17 (1962) 40–41 and 18 (1963) 31–32.

I/2. We define

$$A_i(n) \begin{cases} = \sqrt[i]{i + \sqrt[i+1]{i+1+\cdots+\sqrt[n]{n}}} & \text{for } i \leqslant n \\ = 0 & \text{for } i > n. \end{cases}$$

Then

(1) $$(A_i(n))^i = i + A_{i+1}(n),$$

and $x_k = A_2(k)$.

We shall obtain the desired inequality by a repeated "rationalization" of $x_{n+1} - x_n$, for which we use the identity and inequality

$$x - y = \frac{x^r - y^r}{x^{r-1} + x^{r-2}y + \cdots + y^{r-1}} \leqslant \frac{x^r - y^r}{ry^{r-1}}, \qquad \text{for } x > y > 0.$$

By using this successively for $r = 2, 3, \ldots$, with $x = x_{k+1} = A_2(k+1)$, $y = x_k = A_2(k)$, and (1), we find

$$x_{n+1} - x_n = A_2(n+1) - A_2(n) \leqslant \frac{(A_2(n+1))^2 - (A_2(n))^2}{2A_2(n)}$$

$$= \frac{A_3(n+1) - A_3(n)}{2A_2(n)}$$

$$\leqslant \frac{A_4(n+1) - A_4(n)}{2A_2(n)3(A_3(n))^2} \leqslant \cdots$$

$$\leqslant \frac{\sqrt[n+1]{n+1}}{n!A_2(n)(A_3(n))^2 \cdots (A_n(n))^{n-1}}.$$

The last expression is $< 1/n!$ if and only if

(2) $\qquad (n + 1)^{1/(n+1)} < A_2(n)(A_3(n))^2 \cdots (A_n(n))^{n-1}$.

Since $A_i(n) \geq i^{1/i}$, we have $(A_i(n))^{i-1} \geq i^{(i-1)/i}$. So to establish (2) it suffices to show that

$$2^{1/2} \cdot 3^{2/3} \cdot \ldots \cdot n^{(n-1)/n} > (n + 1)^{1/(n+1)} \quad \text{for } n = 2, 3, \ldots.$$

This is a rather crude inequality, since the factor

$$3^{2/3} > (n + 1)^{1/(n+1)} \qquad \text{for } n = 1, 2, \ldots,$$

and all the other factors on the left are > 1.

I/3. (i) A quadratic function $F(x)$ is determined by three pairs of values $(x, F(x))$. We choose the pairs $(-1, F(-1)), (0, F(0))$ and $(1, F(1))$ and obtain

(1) $\quad F(x) = \dfrac{x(x - 1)}{2} F(-1) - (x + 1)(x - 1) F(0) + \dfrac{(x + 1)x}{2} F(1)$.

Since $|F(-1)|, |F(0)|$ and $|F(1)|$ are ≤ 1, we have

$$2|F(x)| \leq |x(x - 1)| + 2|x^2 - 1| + |x(x + 1)|.$$

In the interval $-1 \leq x \leq 1$, we also have

$$0 \leq 1 + x \leq 2, \qquad 0 \leq 1 - x \leq 2, \qquad \text{and} \quad 0 \leq 1 - x^2 \leq 1,$$

so

$$2|F(x)| \leq |x|\{1 - x + 1 + x\} + 2(1 - x^2) = 2\{|x| + 1 - x^2\},$$

or

$$|F(x)| \leq -\left(|x| - \tfrac{1}{2}\right)^2 + \tfrac{5}{4} \leq \tfrac{5}{4}.$$

To see that this inequality is sharp, consider

$$F(x) = 1 + x - x^2 \quad \text{in } [-1, 1]; \qquad F(\tfrac{1}{2}) = \tfrac{5}{4}.$$

(ii) If $F(x) = ax^2 + bx + c$, then

$$x^2 F\left(\frac{1}{x}\right) = x^2\left(\frac{a}{x^2} + \frac{b}{x} + c\right) = cx^2 + bx + a = G(x).$$

Now by (1),

$$F\left(\frac{1}{x}\right) = \frac{1}{2x^2}[(1 - x)F(-1) + 2(1 - x^2)F(0) + (1 + x)F(1)],$$

so for $-1 \leq x \leq 1$

$$2|G(x)| \leq |1 - x| + 2|1 - x^2| + |1 + x| = 4 - 2x^2,$$

$$|G(x)| \leq 2 - x^2 \leq 2.$$

This inequality is also sharp; for

$$F(x) = 2x^2 - 1, \qquad G(x) = 2 - x^2, \qquad G(0) = 2.$$

I/4. Let $y_i = \dfrac{x_i^2}{x_{i+1}x_{i+2}}$, $i = 1, 2, \ldots, n$, and $x_{n+i} = x_i$. We note that

$y_1 y_2 \cdots y_n = 1$, and write

$$\frac{x_i^2}{x_i^2 + x_{i+1}x_{i+2}} = \frac{x_i^2 + x_{i+1}x_{i+2} - x_{i+1}x_{i+2}}{x_i^2 + x_{i+1}x_{i+2}}$$

$$= 1 - \frac{x_{i+1}x_{i+2}}{x_i^2 + x_{i+1}x_{i+2}} = 1 - \frac{1}{1 + y_i}.$$

Summing from $i = 1$ to $i = n$, we see that the given inequality is equivalent to

$$(1) \qquad \sum_{i=1}^{n} \frac{1}{1 + y_i} \geqslant 1, \qquad y_i > 0, \qquad y_1 y_2 \cdots y_n = 1, \qquad n \geqslant 2.$$

First proof. We have the following identity:

If $y \neq -1$ and $yz = 1$, then $\dfrac{1}{1 + y} + \dfrac{1}{1 + z} = 1.$

If we make a denominator smaller, the corresponding fraction gets larger.

Consequently, if $y > 0$, $z > 0$ and $yz \leqslant 1$, then $\dfrac{1}{1 + y} + \dfrac{1}{1 + z} \geqslant 1.$

We observe now that there is at least one pair y_i, y_j such that $y_i y_j \leqslant 1$ because $(y_1 y_2)(y_2 y_3) \cdots (y_n y_1) = 1$. Those two terms alone make inequality (1) true.

Alternate proof and generalization. If we let $z_i = 1/y_i$, we obtain the equivalent inequality

$$(2) \qquad \frac{1}{1 + z_1} + \frac{1}{1 + z_2} + \cdots + \frac{1}{1 + z_n} \leqslant n - 1,$$

$$z_1 z_2 \cdots z_n = 1 \text{ and } z_i > 0, \; n \geqslant 2.$$

Combining (1) and (2), we have

$$(3) \qquad n - 1 \geqslant \frac{1}{1 + y_1} + \frac{1}{1 + y_2} + \cdots + \frac{1}{1 + y_n} \geqslant 1,$$

$$y_1 y_2 \cdots y_n = 1 \text{ and } y_k > 0.$$

Now let $2t_k = \ln y_k$, then $y_k = e^{2t_k}$ and (3) becomes

$$(4) \qquad n - 1 \geqslant \sum_{k=1}^{n} \frac{1}{1 + e^{2t_k}} \geqslant 1, \qquad \sum_{k=1}^{n} t_k = 0.$$

By using $\tanh t = (e^{2t} - 1)/(e^{2t} + 1)$, changing n to $n + 1$, and replacing t_{n+1} by $-(t_1 + t_2 + \cdots + t_n)$, we can put (4) into the form

(5a) $\tanh(t_1 + t_2 + \cdots + t_n) - 1$

$$\geq (\tanh t_1 - 1) + (\tanh t_2 - 1) + \cdots + (\tanh t_n - 1),$$

(5b) $\tanh(t_1 + t_2 + \cdots + t_n) + 1$

$$\leq (\tanh t_1 + 1) + (\tanh t_2 + 1) + \cdots + (\tanh t_n + 1).$$

A function $F(x)$ satisfying $F(x + y) \geq F(x) + F(y)$ is called *superadditive*; from this definition it follows that

$$F(x_1 + x_2 + \cdots + x_n) \geq F(x_1) + F(x_2) + \cdots + F(x_n).$$

If $-G(x)$ is superadditive, we also say $G(x)$ is *subadditive*. Thus (5a) and (5b) assert that $\tanh t - 1$ is superadditive and that $\tanh t + 1$ is subadditive. The graph of $y = \tanh x$ is given in the figure:

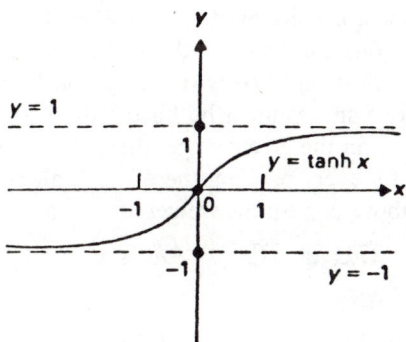

We now show more generally that if $H(x)$ is a non-decreasing function in $(-\infty, \infty)$, $H(0) = 0$, $\lim_{x \to \infty} H(x) = c > 0$ and $\lim_{x \to -\infty} H(x) = -c$, then $H(x) - c$ is superadditive and $H(x) + c$ is subadditive. To show $H(x) - c$ is superadditive, we have to prove that $H(x + y) + c \geq H(x) + H(y)$. We can assume that $x \geq y$. The proof follows easily if we consider each of the three cases:

(i) $x \geq y \geq 0$,   (ii) $x \geq 0 \geq y$,   (iii) $0 \geq x \geq y$.

A similar proof establishes that $H(x) + c$ is subadditive.

Combinatorics and Probability

C / 1. Let T_n be the number of ways of filling up a $2 \times 2 \times n$ hole with $1 \times 1 \times 2$ bricks. We obtain a recursion relation for T_n by expressing it in terms of the numbers of ways of filling smaller holes.

Assume that the long axis of the hole is vertical. The other two edge directions will be called to the right and forward.

First we count the number of ways of filling up the hole if the bottom layer consists of two bricks lying on their sides. They can lie in two ways, with their long axes in the left-right or backward-forward position. The number of these packings is $2T_{n-1}$.

Next consider the packings in which some of the bricks of the bottom layer stick out, but the bricks in layers 1 & 2 fill this sub-box with nothing sticking out. There are 5 such packings: one in which all four bricks are vertical, one in which the two bricks in front are vertical and the two in the back are horizontal, and three more packings of this sort in which the two standing bricks are on the left, in the back, and on the right. This gives $5T_{n-2}$ more packings.

Next let k be an integer, $2 < k \leq n$. We count the number of ways to fill our hole in such a manner that the lowest bottom sub-hole which is completely filled with nothing sticking out has k layers. The bottom layer of such a packing must consist of two vertical and one horizontal bricks. There is no further choice until we get above the k-th layer: the two empty spaces in each layer must be filled with two vertical bricks until we reach layer k, when the two spaces must be filled with a horizontal brick. Thus for all the values of k in the given range, there are 4 ways of filling the k lowest layers. Since for every packing there is one and only one value of k which satisfies the above conditions we get

(1) $$T_n = 2T_{n-1} + 5T_{n-2} + 4T_{n-3} + \cdots + 4T_0.$$

Here we set $T_0 = 1$. Also

(2) $$T_{n-1} = 2T_{n-2} + 5T_{n-3} + 4T_{n-4} + \cdots + 4T_0.$$

Combining (1) and (2), we get

(3) $$T_n = 3T_{n-1} + 3T_{n-2} - T_{n-3}.$$

The characteristic equation† of the latter recurrence relation is

$$x^3 - 3x^2 - 3x + 1 = 0;$$

its roots are -1, $2 + \sqrt{3}$, $2 - \sqrt{3}$. Thus, T_n has the form

$$T_n = c_1(-1)^n + c_2(2 + \sqrt{3})^n + c_3(2 - \sqrt{3})^n.$$

Since the initial values are $T_0 = 1$, $T_1 = 2$, $T_2 = 9$,

$$T_n = (-1)^n/3 + (2 + \sqrt{3})^{n+1}/6 + (2 - \sqrt{3})^{n+1}/6.$$

In particular, $T_{12} = 1/3 + (2 + \sqrt{3})^{13} + (2 - \sqrt{3})^{13} =$ the nearest integer to $(2 + \sqrt{3})^{13} = 4,541,161$.

†See 2nd solution of IMO 63/4 in NML vol. 27, p. 61.

C / 2. Denote by m_j the number of P's containing a_j, $j = 1, 2, \ldots, n$. Since each pair consists of two distinct elements,

$$(1) \qquad\qquad m_1 + m_2 + \cdots + m_n = 2n.$$

The number of pairs $\{P_i, P_j\}$ both containing a_k is $\binom{m_k}{2}$. Now consider the one-to-one mapping $(P_i, P_j) \leftrightarrow (a_i, a_j)$. By hypothesis, (a_i, a_j) is a pair P_k if and only if $P_i \cap P_j \neq \varnothing$. Since there are only n pairs $P_1, P_2, \ldots, P_n$, it follows that

$$(2) \qquad\qquad \sum \binom{m_k}{2} = \frac{1}{2}\left(\sum m_k^2 - \sum m_k \right) \leqslant n$$

(the sums here and subsequently are from $k = 1$ to n). By the power mean†- or Cauchy inequality,

$$\left(\sum m_k \right)^2 \leqslant n \sum m_k^2;$$

using this and (1) and (2), we conclude that

$$(3) \qquad\qquad n \leqslant \frac{1}{2}\left(\sum m_k^2 - \sum m_k \right) \leqslant n.$$

Thus equality must hold in the power mean inequality. But this occurs if and only if all the m_k are equal; by (1), $m_1 = m_2 = \cdots = m_n = 2$.

C / 3. Since our solution of this problem uses graph theory, we introduce the relevant terms about graphs. Although this material is easily accessible (e.g. see [34, Problem 5-3, pp. 35, 249–250]), we include it for the reader's convenience.

A graph is said to be *bipartite* if its vertices can be divided into two classes D and E so that all edges connect a vertex in D and a vertex in E. It is rather obvious that a bipartite graph has no odd circuit (circuit of odd length).

The converse is that, if a graph has no odd circuit, then it is bipartite. Before we show this, let us observe that it suffices to exclude odd simple circuits, i.e. circuits in which no vertex occurs twice. Indeed, if we have an odd circuit in which a vertex v occurs twice, one of the segments between its two occurrences is a shorter odd circuit. If this is still not simple, it will contain an even shorter odd circuit. Ultimately we obtain a simple odd circuit. So if a graph has no simple odd circuit, it has no odd circuit.

If a graph has no odd circuit, then the lengths of all paths connecting any two vertices a, b have the same parity. Otherwise, going from a to b along a path of odd length and returning along a path of even length would give an odd circuit. A partition of the vertex set of the graph into two sets

†See Glossary.

D and E can therefore be obtained thus: In each connected component of the graph, select one vertex arbitrarily. Assign the vertices at an even distance from the vertex chosen in its component to D and the vertices at an odd distance from it to E. If we had an edge connecting two vertices b, c of D, then going from the chosen vertex to b along the shortest path and then along this edge to c would give a path of odd length from the chosen vertex to c. Similarly, there can be no edge connecting points of E.

The *complete graph* K_m is the graph with m vertices in which all pairs of vertices are connected.

We are going to prove the following:

If the set of edges of K_m is partitioned into n sets $A_1, \ldots, A_n$ and $m > 2^n$ then at least one of the A_i contains an odd circuit.

Clearly $m = 1983$ and $n = 10$ yield a solution of the original problem.

For $n = 1$ the result is obvious. We assume the theorem is true for $n - 1$ and prove it for n. If A_n is not bipartite, then it has an odd circuit. So suppose A_n is bipartite.

Let D and E be a partition of the vertices such that no edge of A_n connects two vertices in D or E. One of the two sets, say D, will contain more than 2^{n-1} vertices, and since all the edges connecting vertices of D belong to one of the $n - 1$ sets $A_1, \ldots, A_{n-1}$, one of these must contain an odd circuit by the induction hypothesis.

If m is only 2^n, then the statement need not hold. The example for that is obtained by numbering the vertices of K_m with n binary digits and assigning an edge to A_i if the numbers of its endnodes first differ in the i-th digits.

C / 4. Each time two balls are taken out and one ball is replaced, the number of white balls in the box either decreases by two or else remains the same. Therefore, if p is even, the last ball cannot be white and the probability is zero. If p is odd, the last ball must be white and the probability is one.

C / 5. Letting n be the expected number of tosses, we consider the following three mutually exclusive cases, which cover all the possibilities:

(1) With probability $1/2$, the first toss is a tail. The expected number of tosses is an additional n.

(2) With probability $1/4$, the first two tosses are heads. The expected number of tosses is an additional n.

(3) With probability $1/4$, the first toss is a head and the second toss is a tail. In this case, there will be no more tosses.
Whence,

$$n = \frac{n + 1}{2} + \frac{n + 2}{4} + \frac{2}{4} \qquad \text{or} \qquad n = 6.$$

Appendix A

To provide detailed information regarding the operation of the I.M.O., I give the complete set of instructions which were sent by Czechoslovakia to each of the invited countries. Except for dates, these instructions are essentially the same from year to year.†

25th International Mathematical Olympiad
Regulations

I.   General Provisions

1. The 25th International Mathematical Olympiad (hereinafter I.M.O.) is an international contest for secondary school students in the solution of mathematical problems. It will consist of the contest proper and of related events concerning the problem of detecting and developing pupils with a gift for mathematics.

2. The 25th I.M.O. will be held in the Czechoslovak Socialist Republic from June 29 to July 10, 1984.

3. The related events will be prepared by the I.M.O. organizers. They will include a Symposium and an Exhibition related to the problem area of work with pupils with a talent for mathematics.

4. The contest consists of two papers which will be written in the morning of July 4 and of July 5, 1984. In each of these papers the participants solve three problems in 4–$4\frac{1}{2}$ hours of working time.

5. Each country which accepts the invitation to participate in the 25th I.M.O. may send a delegation consisting of the leader of the delegation, his deputy and six contestants.

---

†This is part of a report on the 25th I.M.O. reprinted from my "Olympiad Corner" in Crux Mathematicorum 10 (1948) 249–261.

6. The I.M.O. is open only to secondary school students or to pupils of schools on an equal level, born after July 5, 1964.

7. The Head (leader) of the delegation and his deputy should be mathematicians or mathematics teachers. They must be able to express themselves exactly and clearly on mathematical and technical aspects of the contest in at least one of the official languages of the I.M.O., i.e., in English, French, German, or Russian.

8. Each country which accepts the invitation to participate in the 25th I.M.O. is requested to send to the organizers 3 to 5 problem proposals with solutions, formulated in one of the official languages of the Olympiad, by April 15. The problems should come from various areas of mathematics, such as are included in math curricula at secondary schools. The solution of these problems should, however, require exceptional mathematical ability and excellent mathematical knowledge on the part of the contestants. It is assumed that the problems will be original and not yet published anywhere. Czechoslovakia, as the organizing country, does not submit any problems for the contest.

9. The Symposium will discuss "Forms and methods of identification and further specialist guidance of pupils with mathematical talent, and pupils' mathematical contests". The Symposium will be attended by the leaders of the delegations and by selected Czechoslovak specialists. The participants are kindly requested to send in their papers and communications in writing to the organizers, in any one of the official languages of the Olympiad, by May 15.

10. The content of the Exhibition will be information on pupils' contests in the solution of mathematical problems (mathematical olympiads, etc.) organized in the countries taking part in the 25th I.M.O., examples of contest problems, and publications intended for the participants in these contests and aimed at widening their mathematical knowledge and developing their talent. Countries taking part in the 25th I.M.O. are requested to send their contributions to the Exhibition to the organizers by May 15.

11. The 25th I.M.O. is managed by the Organizing Committee. The pupils' contest itself is directed by the International Jury of the contest. The Symposium and the Exhibition are organized by the respective commissions of the Organizing Committee of the 25th I.M.O.

12. All expenses related to the stay of the delegations of the participating countries in accordance with the Programme of the 25th I.M.O. will be covered by the organizing country. The participating countries will cover the costs of travel of the members of their delegations to Prague and back,

and the costs of their possible stay in Czechoslovakia before the set day of arrival and after the set day of departure. The organizing country does not cover the costs related to the stay of any other persons.

II.   The International Jury of the 25th I.M.O. and its tasks.

1. The International Jury of the 25th I.M.O. consists of a Chairman who is appointed from the ranks of Czechoslovak specialists by the Ministry of Education of the Czech Socialist Republic in agreement with the Ministry of Education of the Slovak Socialist Republic, and of the leaders of the delegations of the countries participating in the Olympiad. All above said persons have a vote on the Jury's deliberations. Taking part in the work of the Jury is the Deputy Chairman who, in case of the absence of the Chairman, presides over the Jury and takes the vote in his place. After the problems of the second day have been set to the contestants, the members of the Jury are joined in their work by the Deputy Heads of their delegations who only have the right to vote in case of the absence of the leader of their delegation. Also partaking of the work of the Jury, without a vote, is the Chairman of the Problem Selection Committee of the Organizing Committee of the 25th I.M.O., the Main Coordinators (see Article III.7), and possibly other specialists invited by the Chairman in case the situation requires their presence.

2. The Jury may only take decisions on questions that are related to the preparation and realization of the contest proper and to the evaluation of its results in accordance with the provisions of these Regulations. The quorum of the Jury is the presence of at least 50% of the members with the right of vote. The decisions of the Jury are adopted by simple majority. In case of a parity of votes, the vote is decided by the vote of the Chairman.

3. The Jury deliberations are held in the official languages of the Olympiad. In case of necessity, interpreters will attend the meeting without the right of vote.

4. The Jury will start its work at the set date well in advance of the start of the contest in such a manner as to be able to carry out the following preparatory work:

(a) From the preliminary broader selection of problems prepared by the Problem Selection Committee, select 6 problems for the contest.

(b) Determine the sequence of the problems for the contest and their division under the two days of the event.

(c) With regard to the difficulty of the contest problems, determine the duration of the working time and the number of points to be gained for the complete solution of the individual contest problems in such a manner as to make the sum of the points for the complete solution of all 6 problems a total of 40 points.

(d) Prepare and approve the formulation of the texts of the selected contest problems for the 25th I.M.O. in all official languages of the Olympiad.

(e) Decide on possible objections to the translations of the contest problems made by the leaders of the delegations into the working languages of the pupils. The responsibility for the correctness of the translation rests with the delegation leader.

(f) Determine the layout of the sheets with the texts of the contest problems to be distributed to the contestants.

5. Before the start of the contest and during the contest, the Jury will fulfill the following tasks:

(a) On the arrival of the contestants, verify that they satisfy the conditions set for participation under Article I.6. The Jury is entitled to exclude from the contest any pupil who does not meet these conditions.

(b) Decide on possible objections as to the regularity of the contest.

(c) In the course of the contest, decide in each individual case on the answer to the written queries of the contestants related to the text of the contest problems (see Article III.6).

6. Concerning the marking of the solutions to the contest problems, the Jury fulfills the following tasks:

(a) Prior to the coordination of the marking of the solutions, meet with the coordinators to determine the principles for the marking of the solutions of the individual contest problems.

(b) Decide who of the members of the Jury will coordinate the marking of the solutions presented by the Czechoslovak contestants (see Article III.8).

(c) Decide on the marking of the solution of the problems in those cases where agreement could not be reached between the coordinators and the delegation leader.

(d) Agree on the final approval of the marking of the solutions and of the results of the contest.

7. At its final meeting, determine the number of points needed to win the awarded prizes, and decide on the possible awarding of special awards for outstanding original solutions to the individual contest problems.

8. The Chairman of the Jury has to direct the deliberations of the Jury so that the Jury takes an unambiguous decision on every question related to the contest. In case it is necessary to take a vote on any question, the

Chairman must make sure, prior to the vote, that the subject of the vote is clear to all members of the Jury.

III.   Contest problems, written work, and marking of solutions

1. From the draft problems submitted by the participating countries before the given deadline (see Article I.8), the Problem Selection Committee will prepare two variants of 6 contest problems each, and 8 substitute problems. Their texts in the official languages of the Olympiad will be handed over to the members of the jury by the Chairman of the Jury together with the Programme of the 25th I.M.O. on the first day of the Olympiad (June 29).

2. The Chairman of the Jury directs the deliberations of the Jury on the selection of the contest problems in such a manner as to ensure, if possible, agreement of all members of the Jury.

3. The translation of the texts of contest problems from the texts officially approved by the Jury into the working languages of the pupils taking part in the contest is made by the leader of the respective delegation; he is also responsible for the preparation of an adequate number of copies of the texts for the pupils.

4. All persons who know the contest problems are obliged to keep them secret up to the termination of the respective part of the contest.

5. Each contestant writes the solution of the contest problems in his language on a sheet of paper provided by the Organizing Committee of the 25th I.M.O. All other aids—writing utensils, rulers, compasses—the contestants bring with them. The use of other aids outside of those given above is prohibited.

6. The leaders of delegations may answer pupils' queries related to the texts of the contest problems only with the approval of the Jury. The question should be submitted in writing within 30 minutes after the start of the session and the answer will be given to the contestant in writing.

7. The first evaluation of the solutions of the contest problems will be made by the leaders of delegations and their deputies. The coordination of the marking of the solutions of the individual problems is the task of groups of coordinators appointed by the Organizing Committee of the 25th I.M.O. One Main Coordinator and at least two further Coordinators are appointed for each contest problem. They are Czechoslovak mathematicians with appropriate experience in the evaluation of the solutions of mathematical

problems. In each group there is, for each official language, at least one Coordinator who speaks that language.

8. The marking of the solutions submitted by the Czechoslovak participants will be coordinated by the delegation leaders appointed by the Jury (see Article II.6b) in the presence of the main coordinator for the respective problem.

9. The coordination of the marking will follow a time schedule announced by the Chairman of the Jury not later than one day before the start of the coordination.

10. At the coordination of the marking of the solutions of the problems, the leader of the respective delegation or his deputy will, at the request of the Coordinators, translate parts of the solutions or the entire solutions submitted by his contestants into one of the official languages of the Olympiad. The translation may be oral or in writing.

11. The responsibility for the objective evaluation of the problem solutions rests with the Main Coordinator for the respective problem. A protocol will be made on the results of the coordination of the marking of each problem. The protocol will be signed by the leader of the respective delegation and the Main Coordinator for the respective problem.

IV.   Awarding of prizes and conclusion of the 25th I.M.O.

1. Each participant will receive a Diploma certifying to his participation in the Olympiad.

2. The most successful contestants will be awarded 1st, 2nd, and 3rd prizes. Special awards may be presented for outstanding original solutions.

3. The total number of awarded prizes will not exceed half of the number of all contestants. The number of 1st, 2nd, and 3rd prizes awarded will if possible be in the ratio $1 : 2 : 3$.

4. Proposals for special awards for outstanding original solutions of individual problems are submitted to the final meeting of the Jury by the Main Coordinators of the individual problems.

5. The prizes and special awards are not linked with the right to any financial remuneration or material award.

6. The results of the 25th I.M.O. are announced, and the prizes and special awards are handed over, at a solemn public assembly. The diplomas are handed to the participants by representatives of the Ministries of

Education of the Czech Socialist Republic and of the Slovak Socialist Republic and by the Chairman of the Jury.

V.   Written materials of the 25th I.M.O.

Prior to their departure from the venue of the Olympiad, the leaders of delegations and their deputies will receive the draft of selected contest problems, the programme (see Article III.1), the results of the contest as approved by the Jury, a list of contestants who have won prizes and awards at the 25th I.M.O., and the drafts of all problems and their solutions sent in by the participating countries to the organizers of the contest within the set deadline (see Article I.8).

Appendix B

The following tables give a summary of the results of all the IMO's from 1959 through 1985 (there was no IMO in 1980).

1st IMO (1959) in Rumania

Rank	Country	Score (max 320)	Prizes 1st	2nd	3rd	Team Size
1	Rumania	249	1	2	2	8
2	Hungary	233	1	1	2	8
3	Czechoslovakia	192	1	—	—	8
4	Bulgaria	131	—	—	—	8
5	Poland	122	—	—	—	8
6	U.S.S.R.	111	—	—	1	4
7	East Germany	40	—	—	—	8

2nd IMO (1960) in Rumania

Rank	Country	Score (max 360)	Prizes 1st	2nd	3rd	Size
1	Czechoslovakia	257	1	1	2	8
2, 3	Hungary	248	2	2	—	8
2, 3	Rumania	248	1	1	1	8
4	Bulgaria	175	—	—	1	8
5	East Germany	38	—	—	—	8

3rd IMO (1961) in Hungary

Rank	Country	Score (max 320)	Prizes 1st	2nd	3rd	Team Size
1	Hungary	270	2	3	1	8
2	Poland	203	1	—	—	8
3	Rumania	197	—	1	1	8
4	Czechoslovakia	159	—	—	1	8
5	East Germany	146	—	—	1	8
6	Bulgaria	108	—	—	—	8

4th IMO (1962) in Czechoslovakia

Rank	Country	Score (max 368)	1st	2nd	3rd	Team Size
1	Hungary	289	2	3	2	8
2	U.S.S.R.	263	2	2	2	8
3	Rumania	257	—	3	3	8
4, 5	Poland	212	—	1	3	8
4, 5	Czechoslovakia	212	—	1	3	8
6	Bulgaria	196	—	1	2	8
7	East Germany	153	—	1	—	8

5th IMO (1963) in Poland

Rank	Country	Score (max 320)	1st	2nd	3rd	Team Size
1	U.S.S.R.	271	4	3	1	8
2	Hungary	234	—	5	3	8
3	Rumania	191	1	1	3	8
4	Yugoslavia	162	1	2	1	8
5	Czechoslovakia	151	1	—	1	8
6	Bulgaria	145	—	—	3	8
7	East Germany	140	—	—	3	8
8	Poland	134	—	—	2	8

6th IMO (1964) in U.S.S.R.

Rank	Country	Score (max 336)	1st	2nd	3rd	Team Size
1	U.S.S.R.	269	3	1	3	8
2	Hungary	253	3	1	1	8
3	Rumania	213	—	2	3	8
4	Poland	209	1	1	3	8
5	Bulgaria	198	—	—	3	8
6	East Germany	196	—	1	2	8
7	Czechoslovakia	194	—	2	2	8
8	Mongolia	169	—	—	1	8
9	Yugoslavia	155	—	1	1	8

7th IMO (1965) in East Germany

Rank	Country	Score (max 320)	1st	2nd	3rd	Team Size
1	U.S.S.R.	281	5	2	—	8
2	Hungary	244	3	2	2	8
3	Rumania	222	—	4	3	8
4	Poland	178	—	1	3	8
5	East Germany	175	—	2	3	8
6	Czechoslovakia	159	—	1	3	8
7	Yugoslavia	137	—	—	2	8
8	Bulgaria	93	—	—	1	8
9	Mongolia	63	—	—	—	8
10	Finland	62	—	—	—	8

8th IMO (1966) in Bulgaria

Rank	Country	Score (max 320)	1st	2nd	3rd	Team Size
1	U.S.S.R.	293	5	1	1	8
2	Hungary	281	3	1	2	8
3	East Germany	280	3	3	—	8
4	Poland	269	1	4	1	8
5	Rumania	257	1	2	2	8
6	Bulgaria	236	—	1	3	8
7	Yugoslavia	224	—	2	1	8
8	Czechoslovakia	215	—	1	2	8
9	Mongolia	90	—	—	—	8

9th IMO (1967) in Yugoslavia

Rank	Country	Score (max 336)	1st	2nd	3rd	Team Size
1	U.S.S.R.	275	3	3	2	8
2	East Germany	257	3	3	1	8
3	Hungary	251	2	3	3	8
4	Great Britain	231	1	2	4	8
5	Rumania	214	1	1	4	8
6, 7	Bulgaria	159	1	—	1	8
6, 7	Czechoslovakia	159	—	1	3	8
8	Yugoslavia	136	—	—	3	8
9	Sweden	135	—	—	2	8
10	Italy	110	—	1	1	6
11	Poland	101	—	—	1	8
12	Mongolia	87	—	—	1	8
13	France	41	—	—	—	5

10th IMO (1968) in U.S.S.R.

Rank	Country	Score (max 320)	Prizes 1st	Prizes 2nd	Prizes 3rd	Team Size
1	East Germany	304	5	3	—	8
2	U.S.S.R.	298	5	1	2	8
3	Hungary	291	3	3	2	8
4	Great Britain	263	3	2	2	8
5	Poland	262	2	3	2	8
6	Sweden	256	1	2	5	8
7	Czechoslovakia	248	2	4	—	8
8	Rumania	208	1	1	2	8
9	Bulgaria	204	—	3	1	8
10	Yugoslavia	177	—	—	3	8
11	Italy	132	—	—	1	8
12	Mongolia	74	—	—	—	8

11th IMO (1969) in Rumania

Rank	Country	Score (max 320)	Prizes 1st	Prizes 2nd	Prizes 3rd	Team Size
1	Hungary	247	1	4	2	8
2	East Germany	240	—	4	4	8
3	U.S.S.R.	231	1	3	3	8
4	Rumania	219	—	4	2	8
5	Great Britain	193	1	1	1	8
6	Bulgaria	189	—	—	3	8
7	Yugoslavia	181	—	2	2	8
8	Czechoslovakia	170	—	—	3	8
9	Mongolia	120	—	—	1	8
10, 11	France	119	—	1	—	8
10, 11	Poland	119	—	1	—	8
12	Sweden	104	—	—	—	8
13	Belgium	57	—	—	—	8
14	Netherlands	51	—	—	—	8

12th IMO (1970) in Hungary

Rank	Country	Score (max 320)	Prizes 1st	2nd	3rd	Team Size
1	Hungary	233	3	1	3	8
2, 3	East Germany	221	1	2	4	8
2, 3	U.S.S.R.	221	2	1	3	8
4	Yugoslavia	209	—	3	3	8
5	Rumania	208	—	3	4	8
6	Great Britain	180	1	—	6	8
7, 8	Bulgaria	145	—	—	3	8
7, 8	Czechoslovakia	145	—	—	4	8
9	France	141	—	1	4	8
10	Sweden	110	—	—	2	8
11	Poland	105	—	—	1	8
12	Austria	104	—	—	1	8
13	Netherlands	87	—	—	1	8
14	Mongolia	58	—	—	1	8

13th IMO (1971) in Czechoslovakia

Rank	Country	Score (max 320)	Prizes 1st	2nd	3rd	Team Size
1	Hungary	255	4	4	—	8
2	U.S.S.R.	205	1	5	2	8
3	East Germany	142	1	1	4	8
4	Poland	118	1	—	4	8
5, 6	Great Britain	110	—	1	4	8
5, 6	Rumania	110	—	1	4	8
7	Austria	82	—	—	4	8
8	Yugoslavia	71	—	—	2	8
9	Czechoslovakia	55	—	—	1	8
10	Netherlands	48	—	—	2	8
11	Sweden	43	—	—	2	7
12	Bulgaria	39	—	—	—	8
13	France	38	—	—	—	8
14	Mongolia	26	—	—	—	8
15	Cuba	9	—	—	—	4

14th IMO (1972) in Poland

Rank	Country	Score (max 320)	Prizes 1st	Prizes 2nd	Prizes 3rd	Team Size
1	U.S.S.R.	270	2	4	2	8
2	Hungary	263	3	3	2	8
3	East Germany	239	1	3	4	8
4	Rumania	206	1	3	1	8
5	Great Britain	179	—	2	4	8
6	Poland	160	1	1	1	8
7, 8	Austria	136	—	—	5	8
7, 8	Yugoslavia	136	—	—	3	8
9	Czechoslovakia	130	—	—	4	8
10	Bulgaria	120	—	—	2	8
11	Sweden	60	—	—	2	8
12	Netherlands	51	—	—	—	8
13	Mongolia	49	—	—	—	8
14	Cuba	14	—	—	—	3

15th IMO (1973) in U.S.S.R.

Rank	Country	Score (max 320)	Prizes 1st	Prizes 2nd	Prizes 3rd	Team Size
1	U.S.S.R.	254	3	2	3	8
2	Hungary	215	1	2	5	8
3	East Germany	188	—	3	4	8
4	Poland	174	—	2	4	8
5	Great Britain	164	1	—	5	8
6	France	153	—	3	1	8
7	Czechoslovakia	149	—	1	4	8
8	Austria	144	—	—	6	8
9	Rumania	141	—	1	3	8
10	Yugoslavia	137	—	—	5	8
11	Sweden	99	—	1	1	8
12, 13	Bulgaria	96	—	—	1	8
12, 13	Netherlands	96	—	—	2	8
14	Finland	86	—	—	2	8
15	Mongolia	65	—	—	1	8
16	Cuba	42	—	—	1	5

16th IMO (1974) in East Germany

Rank	Country	Score (max 320)	1st	2nd	3rd	Team Size
1	U.S.S.R.	256	2	3	2	8
2	U.S.A.	243	—	5	3	8
3	Hungary	237	1	3	3	8
4	East Germany	236	—	5	2	8
5	Yugoslavia	216	2	1	2	8
6	Austria	212	1	1	4	8
7	Rumania	199	1	1	3	8
8	France	194	1	1	3	8
9	Great Britain	188	—	1	3	8
10	Sweden	187	1	1	—	8
11	Bulgaria	171	—	1	4	8
12	Czechoslovakia	158	—	—	2	8
13	Vietnam	146	1	1	2	5
14	Poland	138	—	—	2	8
15	Netherlands	112	—	—	1	8
16	Finland	111	—	—	1	8
17	Cuba	65	—	—	—	7
18	Mongolia	60	—	—	—	8

17th IMO (1975) in Bulgaria

Rank	Country	Score (max 320)	1st	2nd	3rd	Team Size
1	Hungary	258	—	5	3	8
2	East Germany	249	—	4	4	8
3	U.S.A.	247	3	1	3	8
4	U.S.S.R.	246	1	3	4	8
5	Great Britain	239	2	2	3	8
6	Austria	192	1	1	2	8
7	Bulgaria	186	—	1	4	8
8	Rumania	180	—	1	3	8
9	France	176	1	1	1	8
10	Vietnam	175	—	1	3	7
11	Yugoslavia	163	—	1	1	8
12	Czechoslovakia	162	—	—	2	8
13	Sweden	160	—	2	—	8
14	Poland	124	—	1	1	8
15	Greece	95	—	1	—	8
16	Mongolia	75	—	—	1	8
17	Netherlands	67	—	—	1	8

18th IMO (1976) in Austria

Rank	Country	Score (max 320)	1st	2nd	3rd	Team Size
1	U.S.S.R.	250	4	3	1	8
2	Great Britain	214	2	4	1	8
3	U.S.A.	188	1	4	1	8
4	Bulgaria	174	—	2	6	8
5	Austria	167	1	2	5	8
6	France	165	1	3	1	8
7	Hungary	160	—	3	4	8
8	East Germany	142	—	2	3	8
9	Poland	138	—	—	6	8
10	Sweden	120	—	1	3	8
11	Rumania	118	—	1	3	8
12, 13	Czechoslovakia	116	—	1	3	8
12, 13	Yugoslavia	116	—	1	3	8
14	Vietnam	112	—	1	3	8
15	Netherlands	78	—	—	1	8
16	Finland	52	—	—	1	8
17	Greece	50	—	—	—	8
18	Cuba	16	—	—	—	3
19	West Germany		unclassified			2

19th IMO (1977) in Yugoslavia

Rank	Country	Score (max 320)	1st	2nd	3rd	Team Size
1	U.S.A.	202	2	3	1	8
2	U.S.S.R.	192	1	2	4	8
3, 4	Great Britain	190	1	3	3	8
3, 4	Hungary	190	1	3	2	8
5	Netherlands	185	1	2	3	8
6	Bulgaria	172	—	3	3	8
7	West Germany	165	1	1	4	8
8	East Germany	163	2	1	1	8
9	Czechoslovakia	161	—	3	2	8
10	Yugoslavia	159	—	3	3	8
11	Poland	157	1	2	2	8
12	Austria	151	1	1	2	8
13	Sweden	137	1	1	2	8
14	France	126	1	—	—	8
15	Rumania	122	—	1	2	8
16	Finland	88	—	—	1	8
17	Mongolia	49	—	—	—	8
18	Cuba	41	—	—	—	4
19	Belgium	33	—	—	—	7
20	Italy	22	—	—	—	5
21	Algeria	17	—	—	—	3

20th IMO (1978) in Rumania

Rank	Country	Score (max 320)	Prizes 1st	Prizes 2nd	Prizes 3rd	Team Size
1	Rumania	237	2	3	2	8
2	U.S.A.	225	1	3	3	8
3	Great Britain	201	1	2	2	8
4	Vietnam	200	—	2	6	8
5	Czechoslovakia	195	—	2	3	8
6	West Germany	184	1	—	3	8
7	Bulgaria	182	—	1	3	8
8	France	179	—	2	4	8
9	Austria	174	—	3	2	8
10	Yugoslavia	171	—	1	2	8
11	Netherlands	157	—	1	1	8
12	Poland	156	—	—	2	8
13	Finland	118	—	—	2	8
14	Sweden	117	—	—	1	8
15	Cuba	68	—	—	2	4
16	Turkey	66	—	—	—	8
17	Mongolia	61	—	—	—	8

21st IMO (1979) in Great Britain

Rank	Country	Score (max 320)	Prizes 1st	Prizes 2nd	Prizes 3rd	Team Size
1	U.S.S.R.	267	2	4	1	8
2	Rumania	240	1	4	2	8
3	West Germany	235	1	5	2	8
4	Great Britain	218	—	4	4	8
5	U.S.A.	199	1	2	2	8
6	East Germany	180	—	2	2	8
7	Czechoslovakia	178	1	—	4	8
8	Hungary	176	—	2	2	8
9	Yugoslavia	168	—	1	4	8
10	Poland	160	—	2	3	8
11	France	155	1	—	1	8
12	Austria	152	—	—	4	8
13	Bulgaria	150	—	—	5	8
14	Sweden	143	—	2	1	8
15	Vietnam	134	1	3	—	4
16	Netherlands	131	—	1	1	8
17	Israel	119	—	—	2	8
18	Finland	89	—	—	1	8
19	Belgium	66	—	—	1	8
20	Greece	57	—	—	1	8
21	Cuba	35	—	—	—	4
22	Brazil	19	—	—	—	5
23	Luxembourg	7	—	—	—	1

22nd IMO (1981) in U.S.A.

Rank	Country	Score (max 336)	1st	Prizes 2nd	3rd	Team Size
1	United States	314	4	3	1	8
2	West Germany	312	5	2	1	8
3	Great Britain	301	3	4	1	8
4	Austria	290	4	2	1	8
5	Bulgaria	287	2	3	3	8
6	Poland	259	2	3	1	8
7	Canada	249	2	2	1	8
8	Yugoslavia	246	1	2	3	8
9	Soviet Union	230	3	2	1	6
10	Netherlands	219	—	3	1	8
11	France	209	2	—	3	8
12	Sweden	207	—	1	3	8
13	Finland	206	1	1	3	8
14	Czechoslovakia	190	1	3	1	5
15	Israel	175	1	—	3	6
16	Brazil	172	1	—	—	8
17	Hungary	164	3	1	—	4
18	Cuba	141	—	1	—	8
19	Belgium	139	—	2	—	8
20	Romania	136	—	2	2	4
21	Australia	122	—	—	1	8
22	Greece	104	—	—	—	8
23	Columbia	93	—	—	—	8
24	Venezuela	64	—	—	—	8
25	Luxembourg	42	1	—	—	1
26	Tunisia	32	—	—	—	2
27	Mexico	5	—	—	—	5

23rd IMO (1982) in Hungary

Rank	Country	Score (max 168)	1st	2nd	3rd	Team Size
1	West Germany	145	2	2	—	4
2	Soviet Union	137	2	1	1	4
3, 4	East Germany	136	2	1	1	4
3, 4	United States	136	1	2	1	4
5	Vietnam	133	1	2	1	4
6	Hungary	125	—	3	1	4
7	Czechoslovakia	115	—	2	2	4
8	Finland	113	—	2	1	4
9	Bulgaria	108	—	—	4	4
10	Great Britain	103	—	—	4	4
11	Romania	99	—	1	2	4
12	Yugoslavia	98	—	2	—	4
13	Poland	96	—	1	2	4
14	Netherlands	92	—	1	1	4
15	France	89	1	—	—	4
16	Austria	82	1	—	1	4
17	Canada	78	—	—	2	4
18	Israel	75	—	—	1	4
19	Sweden	74	—	—	2	4
20, 21	Australia	66	—	—	1	4
20, 21	Brazil	66	—	—	1	4
22	Mongolia	56	—	—	1	4
23	Greece	55	—	—	—	4
24	Belgium	50	—	—	1	4
25	Cuba	44	—	—	—	4
26	Columbia	34	—	—	—	4
27, 28	Algeria	23	—	—	—	3
27, 28	Venezuela	23	—	—	—	4
29	Tunisia	19	—	—	—	4
30	Kuwait	4	—	—	—	4

24th IMO (1983) in France

Rank	Country	Score (max 252)	Prizes 1st	Prizes 2nd	Prizes 3rd	Team Size
1	West Germany	212	4	1	—	6
2	U.S.A.	171	1	3	2	6
3	Hungary	170	—	4	2	6
4	U.S.S.R.	169	1	3	2	6
5	Romania	161	1	2	3	6
6	Vietnam	148	—	3	3	6
7	Netherlands	143	1	3	—	6
8	Czechoslovakia	142	1	1	3	6
9	Bulgaria	137	—	1	4	6
10	France	123	—	2	3	6
11	Great Britain	121	—	3	1	6
12	Fast Germany	117	—	—	5	6
13	Finland	103	—	—	3	6
14	Canada	102	—	—	4	6
15	Poland	101	—	—	3	6
16	Israel	97	—	—	5	6
17	Greece	96	—	—	3	6
18	Yugoslavia	89	—	—	5	6
19	Australia	86	—	1	2	6
20	Brazil	77	—	—	3	6
21	Sweden	47	—	—	—	6
22	Austria	45	—	—	—	6
23	Spain	37	—	—	—	4
24	Cuba	36	—	—	1	6
25	Morocco	32	—	—	—	6
26	Belgium	31	—	—	—	6
27	Tunisia	26	—	—	—	6
28	Colombia	21	—	—	—	6
29	Luxemburg	13	—	—	—	2
30	Algeria	6	—	—	—	6
31	Kuwait	4	—	—	—	6
32	Italy	2	—	—	—	6

25th IMO (1984) in Czechoslovakia

Rank	Country	Score (max 252)	1st	2nd	3rd	Team Size
1	Soviet Union	235	5	1	0	6
2	Bulgaria	203	2	3	1	6
3	Romania	199	2	2	2	6
4, 5	Hungary	195	1	4	1	6
4, 5	U.S.A.	195	1	4	1	6
6	Great Britain	169	1	3	1	6
7	Vietnam	162	1	2	3	6
8	East Germany	161	1	2	3	6
9	West Germany	150	—	2	4	6
10	Mongolia	146	—	3	2	6
11	Poland	140	—	1	5	6
12	France	126	—	2	2	6
13	Czechoslovakia	125	—	2	2	6
14	Yugoslavia	105	—	—	4	6
15	Australia	103	—	1	2	6
16	Austria	97	—	1	2	6
17	Netherlands	93	—	1	2	6
18	Brazil	92	—	—	3	6
19	Greece	88	—	1	—	6
20	Canada	83	—	—	1	6
21	Colombia	80	—	—	2	6
22	Cuba	67	—	—	1	6
23, 24	Belgium	56	—	—	1	6
23, 24	Morocco	56	—	—	1	6
25	Sweden	53	—	—	—	6
26	Cyprus	47	—	—	1	6
27	Spain	43	—	—	—	6
28	Algeria	36	—	—	—	4
29	Finland	31	—	—	—	6
30	Tunisia	29	—	—	—	6
31	Norway	24	—	—	1	1
32	Luxembourg	22	—	—	1	1
33	Kuwait	9	—	—	—	6
34	Italy	0	—	—	—	6

26th IMO (1985) in Finland

Rank	Country	Score (max 252)	1st	2nd	3rd	Team Size
1	Rumania	201	3	3	—	6
2	U.S.A.	180	2	4	—	6
3	Hungary	168	2	2	2	6
4	Bulgaria	165	2	3	—	6
5	Vietnam	144	1	3	1	6
6	U.S.S.R.	140	1	2	2	6
7	West Germany	139	1	1	4	6
8	East Germany	136	—	3	3	6
9	France	125	—	2	3	6
10	Great Britain	121	—	2	3	6
11	Australia	117	1	1	2	6
12, 13	Canada	105	—	1	4	6
12, 13	Czechoslovakia	105	—	3	1	6
14	Poland	101	—	1	4	6
15	Brazil	83	—	—	2	6
16	Israel	81	—	1	—	6
17	Austria	77	—	—	3	6
18	Cuba	74	—	—	2	6
19	The Netherlands	72	—	—	1	6
20	Greece	69	—	1	1	6
21	Yugoslavia	68	—	—	2	6
22	Sweden	65	—	—	1	6
23	Mongolia	62	—	1	—	6
24, 25	Belgium	60	1	—	1	6
24, 25	Morocco	60	—	—	2	6
26, 27	Colombia	54	—	—	2	6
26, 27	Turkey	54	—	—	2	6
28	Tunisia	46	—	—	2	4
29	Algeria	36	—	—	—	6
30	Norway	34	—	—	—	6
31	Iran	28	—	1	—	1
32, 33	China	27	—	—	1	2
32, 33	Cyprus	27	—	—	1	6
34, 35	Finland	25	—	—	—	6
34, 35	Spain	25	—	—	—	4
36	Italy	20	—	—	—	5
37	Iceland	13	—	—	—	2
38	Kuwait	7	—	—	—	5

List of Symbols

$[ABC]$	area of $\triangle ABC$
$\simeq$ or $\approx$	approximately equal to
$\cong$	congruent (in geometry)
$a \equiv b \pmod{p}$	$a - b$ is divisible by p; *Congruence*, see Glossary
$a \not\equiv b \pmod{p}$	$a - b$ is not divisible by p
$\equiv$	identically equal to
$[x]$ or $\lfloor x \rfloor$	integer part of x, i.e. greatest integer not exceeding x
$\lceil x \rceil$	least integer greater than or equal to x
$\binom{n}{k}, C(n, k)$	binomial coefficient, see Glossary; also the number of combinations of n things, k at a time
(n, k), G.C.D. of n, k	the greatest common divisor of n and k
$p \mid n$	p divides n
$p \nmid n$	p does not divide n
$n!$	n factorial $= 1 \cdot 2 \cdot 3 \cdot \ldots (n-1)n, 0! = 1$
$\prod_{i=1}^{n} a_i$	the product $a_1 \cdot a_2 \cdot \cdots \cdot a_n$
$\sim$	similar in geometry
$\sum_{i=1}^{n} a_i$	the sum $a_1 + a_2 + \cdots + a_n$
$\circ$	$f \circ g(x) = f[g(x)]$, see *Composition* in Glossary
$K_1 \cup K_2$	union of sets K_1, K_2
$K_1 \cap K_2$	intersection of sets K_1, K_2
A. M.	arithmetic mean, see *Mean* in Glossary
G. M.	geometric mean, see *Mean* in Glossary
H. M.	harmonic mean, see *Mean* in Glossary
$[a, b]$	closed interval, i.e. all x such that $a \leqslant x \leqslant b$
(a, b)	open interval, i.e. all x such that $a < x < b$

Glossary of some frequently used terms and theorems.

Arithmetic mean (average). see *Mean*

Arithmetic mean-geometric mean inequality (A.M.-G.M. inequality).
If $a_1, a_2, \ldots, a_n$ are n non-negative numbers, then

$$\frac{1}{n} \sum_{i=1}^{n} a_i \geq \left[\prod_{i=1}^{n} a_i \right]^{1/n} \quad \text{with equality if and only if} \quad a_1 = a_2 = \cdots = a_n.$$

Weighted arithmetic mean-geometric mean inequality.
If, in addition, $w_1, w_2, \ldots, w_n$ are non-negative numbers (weights) whose sum is 1, then

$$\sum_{i=1}^{n} w_i a_i \geq \prod_{i=1}^{n} a_i^{w_i} \quad \text{with equality if and only if} \quad a_1 = a_2 = \cdots = a_n.$$

For a proof, use Jensen's inequality below, applied to $f(x) = -\log x$.

Arithmetic Series. see *Series*

Binomial coefficient.

$$\binom{n}{k} = \frac{n!}{k!(n-k)!} = \binom{n}{n-k} = \text{coefficient of } y^k \text{ in the expansion } (1+y)^n.$$

Also $\binom{n+1}{k+1} = \binom{n}{k+1} + \binom{n}{k}$. (See *Binomial theorem* and List of Symbols.)

Binomial theorem.

$$(x+y)^n = \sum_{k=0}^{n} \binom{n}{k} x^{n-k} y^k, \text{ where}$$

$$\binom{n}{k} = \frac{n(n-1) \cdots (n-k+1)}{1 \cdot 2 \cdot \ldots \cdot (k-1)k} = \frac{n!}{k!(n-k)!}.$$

Cauchy's inequality.

For vectors $\mathbf{x}, \mathbf{y}$, $\quad |\mathbf{x} \cdot \mathbf{y}| \leqslant |\mathbf{x}||\mathbf{y}|$; $\quad$ componentwise, for real numbers $x_i, y_i, \quad i = 1, 2, \ldots, n,$

$$|x_1 y_1 + x_2 y_2 + \cdots + x_n y_n| \leqslant \left[\sum_{i=1}^{n} x_i^2 \right]^{1/2} \left[\sum_{i=1}^{n} y_i^2 \right]^{1/2}.$$

There is equality if and only if $\mathbf{x}, \mathbf{y}$ are collinear, i.e., if and only if $x_i = k y_i, i = 1, 2, \ldots, n$. A proof for vectors follows from the definition of dot product $\mathbf{x} \cdot \mathbf{y} = |\mathbf{x}||\mathbf{y}| \cos(\mathbf{x}, \mathbf{y})$ or by considering the discriminant of the quadratic function $q(t) = \Sigma (y_i t - x_i)^2$.

Centroid of a triangle.

Point of intersection of the medians

Ceva's theorem.

If AD, BE, CF are concurrent cevians of a triangle ABC, then

(i) $\qquad\qquad\qquad BD \cdot CE \cdot AF = DC \cdot EA \cdot FB.$

Conversely, if AD, BE, CF are three cevians of a triangle ABC such that (i) holds, then the three cevians are concurrent. (A *cevian* is a segment joining a vertex of a triangle with a point on the opposite side.)

Chinese remainder theorem.

Let $m_1, m_2, \ldots, m_n$ denote n positive integers that are relatively prime in pairs, and let $a_1, a_2, \ldots, a_n$ denote any n integers. Then the congruences $x \equiv a_i (\bmod m_i), i = 1, 2, \ldots, n$ have common solutions; any two solutions are congruent modulo $m_1 m_2 \ldots m_n$. For a proof, see [112, p. 31].

Circumcenter of $\triangle ABC$.

Center of circumscribed circle of $\triangle ABC$

Circumcircle of $\triangle ABC$.

Circumscribed circle of $\triangle ABC$

Complex numbers.

Numbers of the form $x + iy$, where x, y are real and $i = \sqrt{-1}$.

Composition of functions.

$F(x) = f \circ g(x) = f[g(x)]$ is the composite of functions, f, g, where the range of g is the domain of f.

Congruence.

$a \equiv b (\bmod p)$ *read "a is congruent to b modulo p" means that $a - b$ is divisible by p.*

Concave function.

$f(x)$ is concave if $-f(x)$ is convex; see *Convex function.*

Convex function.

A function $f(x)$ is convex in an interval I if for all x_1, x_2 in I and for all non-negative weights w_1, w_2 with sum 1,

$$w_1 f(x_1) + w_2 f(x_2) \geq f(w_1 x_1 + w_2 x_2).$$

Geometrically this means that the graph of f between $(x_1, f(x_1))$ and $(x_2, f(x_2))$ lies below its secants.

We state the following useful facts:

1. A continuous function which satisfies the above inequality for $w_1 = w_2 = 1/2$ is convex.
2. A twice differentiable function f is convex if and only if $f''(x)$ is non-negative in the interval in question.
3. The graph of a differentiable convex function lies above its tangents.

For an even more useful fact, see *Jensen's inequality*.

Convex hull of a pointset S.

The intersection of all convex sets containing S

Convex pointset.

A pointset S is convex if, for every pair of points P, Q in S, all points of the segment PQ are in S.

Cross product (vector product) $\mathbf{x} \times \mathbf{y}$ *of two vectors. see Vectors.*

Cyclic polygon.

Polygon that can be inscribed in a circle.

de Moivre's theorem.

$(\cos \theta + i \sin \theta)^n = \cos n\theta + i \sin n\theta$. For a proof, see NML vol. 27, p. 49.

Determinant of a square matrix M (det M).

A multi-linear function $f(C_1, C_2, \ldots, C_n)$ of the columns of M with the properties

$$f(C_1, C_2, \ldots, C_i, \ldots, C_j, \ldots, C_n) = -f(C_1, C_2, \ldots, C_j, \ldots, C_i, \ldots, C_n)$$

and $\det I = 1$. Geometrically, $\det(C_1, C_2, \ldots, C_n)$ is the signed volume of the n-dimensional oriented parallelepiped with coterminal side vectors $C_1, C_2, \ldots C_n$.

Dirichlet's principle. see Pigeonhole principle.

Dot product (scalar product) $\mathbf{x} \cdot \mathbf{y}$ *of two vectors. see Vectors.*

Escribed circle. see Excircle.

Euclid's algorithm.

A process of repeated divisions yielding the greatest common divisor of two integers, $m > n$:

$$m = nq_1 + r_1, \quad q_1 = r_1q_2 + r_2, \quad \ldots, \quad q_k = r_kq_{k+1} + r_{k+1};$$

the last non-zero remainder is the GCD of m and n. For a detailed discussion, see e.g. C. D. Olds, *Continued Fractions*, NML vol. 9 (1963), p. 16.

Euler's extension of Fermat's theorem. see *Fermat's theorem.*

Euler's theorem on the distance d between in- and circumcenters of a triangle.
$d = \sqrt{R^2 - 2rR}$, where r, R are the radii of the inscribed and circumscribed circles.

Excircle of $\triangle ABC$.
A circle that touches one side of the triangle internally and the other two (extended) externally.

Fermat's Theorem.
If p is a prime, $a^p \equiv a \pmod{p}$.
Euler's extension of $-$.
If m is relatively prime to n, then $m^{\phi(n)} \equiv 1 \pmod{n}$, where the Euler $\phi(n)$ function is defined to be the number of positive integers $\leq n$ and relatively prime to n. There is a simple formula for ϕ:

$$\phi(n) = n\prod\left(1 - \frac{1}{p_j}\right), \quad \text{where } p_j \text{ are distinct prime factors of } n.$$

Fundamental summation formula.
Our name for the telescoping sum formula, to point out its similarity to the Fundamental Theorem of Calculus. see *Summation of Series.*

Geometric mean. see *Mean, geometric.*

Geometric series. see *Series, geometric.*

Harmonic mean. see *Mean, harmonic.*

Heron's formula.
The area of $\triangle ABC$ with sides a, b, c is
$[ABC] = \sqrt{s(s-a)(s-b)(s-c)}$, where $s = \frac{1}{2}(a + b + c)$.

Hölder's inequality.
If a_i, b_i are non-negative numbers, and if p, q are positive numbers such that $(1/p) + (1/q) = 1$, then
$$a_1b_1 + a_2b_2 + \cdots + a_nb_n$$
$$\leqslant \left(a_1^p + a_2^p + \cdots + a_n^p\right)^{1/p}\left(b_1^q + b_2^q + \cdots + b_n^q\right)^{1/q}$$
with equality if and only if $a_i = kb_i$, $i = 1, 2, \ldots, n$. Cauchy's inequality corresponds to the special case $p = q = 2$.

Homogeneous.

$f(x, y, z, \ldots)$ is homogeneous of degree k if

$$f(tx, ty, tz, \ldots) = t^k f(x, y, z, \ldots).$$

A system of linear equations is called homogeneous if each equation is of the form $f(x, y, z, \ldots) = 0$ with f homogeneous of degree 1.

Homothety.

A dilatation (simple stretch or compression) of the plane (or space) which multiplies all distances from a fixed point, called the *center of homothety* (or *similitude*), by the same factor $\lambda \neq 0$. This mapping (transformation) is a similarity which transforms each line into a parallel line, and the only point unchanged (invariant) is the center. Conversely, if any two similar figures have their corresponding sides parallel, then there is a homothety which transforms one of them into the other, and the center of homothety is the point of concurrence of all lines joining pairs of corresponding points. Two physical examples are a photo enlarger and a pantograph.

Incenter of $\triangle ABC$.

Center of inscribed circle of $\triangle ABC$

Incircle of $\triangle ABC$.

Inscribed circle of $\triangle ABC$

Inequalities.

A.M. − G.M.—see *Arithmetic mean*
A.M. − H.M.—see *Mean, Harmonic*
Cauchy—see *Cauchy's-*
H.M. − G.M.—see *Mean, Harmonic*
Hölder—see *Hölder's-*
Jensen—see *Jensen's-*
Power mean—see *Power mean-*
Schur—see *Schur's-*
Triangle—see *Triangle-*

Inverse function.

$f: X \to Y$ has an inverse f^{-1} if for every y in the range of f there is a unique x in the domain of f such that $f(x) = y$; then $f^{-1}(y) = x$, and $f^{-1} \circ f$, $f \circ f^{-1}$ are the identity functions. See also *Composition*.

Irreducible polynomial.

A polynomial $g(x)$, not identically zero, is irreducible over a field F if there is no factoring, $g(x) = r(x)s(x)$, of $g(x)$ into two polynomials $r(x)$ and $s(x)$ of positive degrees over F. For example, $x^2 + 1$ is irreducible over the real number field, but reducible, $(x + i)(x - i)$, over the complex number field.

Isoperimetric theorem for triangles.
Among all triangles with given area, the equilateral triangle has the smallest perimeter.

Jensen's inequality.
If $f(x)$ is convex in an interval I and $w_1, w_2, \ldots, w_n$ are arbitrary non-negative weights whose sum is 1, then

$$w_1 f(x_1) + w_2 f(x_2) + \cdots + w_n f(x_n)$$
$$\geq f(w_1 x_1 + w_2 x_2 + \cdots + w_n x_n)$$

for all x_i in I.

Matrix.
A rectangular array of number (a_{ij})

Mean of n numbers.

$$\text{Arithmetic mean } (average) = A.M. = \frac{1}{n} \sum_{i=1}^{n} a_i$$

$$\text{Geometric mean} = G.M. = \sqrt[n]{a_1 a_2 \cdots a_n}, \ a_i \geq 0$$

$$\text{Harmonic mean} = H.M. = \left(\frac{1}{n} \sum_{i=1}^{n} \frac{1}{a_i} \right)^{-1}, \ a_i > 0$$

A.M.− G.M.− H.M. inequalities
$A.M. \geq G.M. \geq H.M.$ with equality if and only if all n numbers are equal.

$$\text{Power mean} = P(r) = \left[\frac{1}{n} \sum_{i=1}^{n} a_i^r \right]^{1/r}, \ a_i > 0, \ r \neq 0, \ |r| < \infty$$

$$= [\Pi a_i]^{1/n} \quad \text{if } r = 0$$
$$= \min(a_i) \quad \text{if } r = -\infty$$
$$= \max(a_i) \quad \text{if } r = \infty$$

Special cases: $P(0) = G.M., \ P(-1) = H.M., \ P(1) = A.M.$
It can be shown that $P(r)$ is continuous on $-\infty \leq r \leq \infty$, that is

$$\lim_{r \to 0} P(r) = [\Pi a_i]^{1/n}, \quad \lim_{r \to -\infty} P(r) = \min(a_i),$$

$$\lim_{r \to \infty} P(r) = \max(a_i).$$

− inequality.
$P(r) \leq P(s)$ for $-\infty \leq r < s \leq \infty$, with equality if and only if all the a_i are equal. For a proof, see [108, pp. 76–77.]

Menelaus' theorem.
If D, E, F, respectively, are three collinear points on the sides BC, CA, AB of a triangle ABC, then

(i) $$BD \cdot CE \cdot AF = -DC \cdot EA \cdot FB.$$

Conversely, if D, E, F, respectively are three points on the sides BC, CA, AB of a triangle ABC such that (i) holds, then D, E, F are collinear.

Orthocenter of $\triangle ABC$.
    Point of intersection of altitudes of $\triangle ABC$

Periodic function.
    $f(x)$ is periodic with period a if $f(x + a) = f(x)$ for all x.

Pigeonhole principle (Dirichlet's box principle).
    If n objects are distributed among $k < n$ boxes, some box contains at least two objects.

Polynomial in x of degree n.
    Function of the form $P(x) = \sum_{i=0}^{n} c_i x^i$, $c_n \neq 0$.
    Irreducible-see *Irreducible*-

Radical axis of two non-concentric circles.
    Locus of points of equal powers with respect to the two circles. (If the circles intersect, it is the line containing the common chord.)

Radical center of three circles with non-collinear centers.
    Common intersection of the three radical axes of each pair of circles.

Root of an equation.
    Solution of an equation

Roots of unity.
    Solutions of the equation $x^n - 1 = 0$.

Schur's inequality.
    $x^n(x - y)(x - z) + y^n(y - z)(y - x) + z^n(z - x)(z - y) \geq 0$,   for all real $x, y, z, n \geq 0$. see [108].

Series.
    Arithmetic: $\sum_{j=1}^{n} a_j$ with $a_{j+1} = a_j + d$, d the common difference.
    Geometric: $\sum_{j=0}^{n-1} a_j$ with $a_{j+1} = r a_j$, r the common ratio.
    Summation of—
        Linearity: $\sum_k [aF(k) + bG(k)] = a\sum_k F(k) + b\sum_k G(k)$.
    Fundamental theorem of- or *Telescoping sums theorem*;
    $$\sum_{k=1}^{n} [F(k) - F(k - 1)] = F(n) - F(0)$$
    (so named in analogy with Fundamental Theorem of Calculus).

By choosing F appropriately, we can obtain the following sums:

$$\sum_{k=1}^{n} 1 = n, \quad \sum_{k=1}^{n} k = \frac{1}{2}n(n+1), \quad \sum_{k=1}^{n} k^2 = \frac{1}{6}n(n+1)(2n+1),$$

$$\sum_{k=1}^{n} [k(k+1)]^{-1} = 1 - \frac{1}{n+1},$$

$$\sum_{k=1}^{n} [k(k+1)(k+2)]^{-1} = \frac{1}{4} - \frac{1}{2(n+1)(n+2)}.$$

$$\sum_{k=1}^{n} ar^{k-1} = a(1-r^n)/(1-r), \quad \text{the sum of a geometric series,}$$

see above.

$$\sum_{k=1}^{n} \cos 2kx = \frac{\sin nx \cos(n+1)x}{\sin x},$$

$$\sum_{k=1}^{n} \sin 2kx = \frac{\sin nx \sin(n+1)x}{\sin x}$$

Subadditive.
  A function $f(x)$ is subadditive if $f(x+y) \le f(x) + f(y)$

Superadditive.
  A function $g(x)$ is superadditive if $g(x+y) \ge g(x) + g(y)$

Telescoping sum. see *Series, Fundamental theorem of summation of—*

Trigonometric identities.

$$\left. \begin{array}{l} \sin(x \pm y) = \sin x \cos y \pm \sin y \cos x \\ \cos(x \pm y) = \cos x \cos y \mp \sin x \sin y \end{array} \right\} \quad \text{addition formulas}$$

$$\left. \begin{array}{l} \sin nx = \cos^n x \left\{ \binom{n}{1}\tan x - \binom{n}{3}\tan^3 x + \cdots \right\} \\ \cos nx = \cos^n x \left\{ 1 - \binom{n}{2}\tan^2 x + \binom{n}{4}\tan^4 x - \cdots \right\} \end{array} \right\} \quad \begin{array}{l} \text{consequences of} \\ \text{de Moivre's} \\ \text{theorem} \end{array}$$

$$\sin 2x + \sin 2y + \sin 2z - \sin 2(x+y+z)$$
$$= 4\sin(y+z)\sin(z+x)\sin(x+y),$$

$$\cos 2x + \cos 2y + \cos 2z + \cos 2(x+y+z)$$
$$= 4\cos(y+z)\cos(z+x)\cos(x+y),$$

$$\sin(x+y+z)$$
$$= \cos x \cos y \cos z(\tan x + \tan y + \tan z - \tan x \tan y \tan z),$$

$$\cos(x+y+z)$$
$$= \cos x \cos y \cos z(1 - \tan y \tan z - \tan z \tan x - \tan x \tan y).$$

Vectors
  One may consider an n-dimensional vector as an ordered n-tuple of real numbers: $\mathbf{x} = (x_1, x_2, \ldots, x_n)$. Its product with any real number a is the vector $a\mathbf{x} = (ax_1, ax_2, \ldots, ax_n)$. The *sum* of two vectors $\mathbf{x}$ and $\mathbf{y}$ is the vector $\mathbf{x} + \mathbf{y} = (x_1 + y_1, x_2 + y_2, \ldots, x_n + y_n)$ (parallelogram or triangle law of addition). The *dot-* or *scalar product* $\mathbf{x} \cdot \mathbf{y}$ is

defined geometrically as $|\mathbf{x}||\mathbf{y}|\cos\theta$, where $|\mathbf{x}|$ denotes the length of $\mathbf{x}$, etc., and θ is the angle between the two vectors. Algebraically, the dot product is defined as the number

$$\mathbf{x}\cdot\mathbf{y} = x_1y_1 + x_2y_2 + \cdots + x_ny_n, \quad \text{and}$$

$$|\mathbf{x}|^2 = \mathbf{x}\cdot\mathbf{x} = x_1^2 + x_2^2 + \cdots + x_n^2.$$

In 3-dimensional space E^3, the *vector-* or *cross product* $\mathbf{x}\times\mathbf{y}$ is defined geometrically as a vector orthogonal to both $\mathbf{x}$ and $\mathbf{y}$, whose magnitude is $|\mathbf{x}||\mathbf{y}|\sin\theta$ and directed according to the right hand screw convention. Algebraically, the cross product of $\mathbf{x} = (x_1, x_2, x_3)$ and $\mathbf{y} = (y_1, y_2, y_3)$ is given by the vector

$$\mathbf{x}\times\mathbf{y} = (x_2y_3 - x_3y_2, x_3y_1 - x_1y_3, x_1y_2 - x_2y_1).$$

It follows from the geometric definition that the *triple scalar product* $\mathbf{x}\cdot\mathbf{y}\times\mathbf{z}$ is the signed volume of the oriented parallelepiped having $\mathbf{x}$, $\mathbf{y}$ and $\mathbf{z}$ as coterminal sides. It can conveniently be written as

$$\mathbf{x}\cdot\mathbf{y}\times\mathbf{z} = \begin{vmatrix} x_1 & x_2 & x_3 \\ y_1 & y_2 & y_3 \\ z_1 & z_2 & z_3 \end{vmatrix} = \det(\mathbf{x}, \mathbf{y}, \mathbf{z}), \text{ see also } Determinant.$$

Zero of a function $f(x)$.
    Any point x for which $f(x) = 0$.

References

General:

1. B. Averbach and O. Chein, *Mathematics: Problem Solving through Recreational Mathematics*, Freeman, San Francisco, 1980.
2. W. W. R. Ball and H. S. M. Coxeter, *Mathematical Recreations*, Macmillan, N.Y., 1939.
3. A. Beck, M. N. Bleicher, and D. W. Crowe, *Excursions into Mathematics*, Worth, N.Y., 1969.
4. D. M. Campbell, *The Whole Craft of Number*, Prindle, Weber, and Schmidt, Boston, 1976.
5. R. Courant and H. Robbins, *What is Mathematics?*, Oxford University Press, Oxford, 1941.
6. S. Gudder, *A Mathematical Journey*, McGraw-Hill, N.Y., 1976.
7. D. H. Hilbert and S. Cohn-Vossen, *Geometry and the Imagination*, Chelsea, N.Y., 1952.
8. R. Honsberger, *Mathematical Gems*, The Dolciani Mathematical Expositions, Vols. I, II, IX, M.A.A., Wash., D.C., 1973, 1976, and 1985.
9. R. Honsberger, *Mathematical Morsels*, The Dolciani Mathematical Expositions, Vol. III, M.A.A., Wash., D.C., 1978.
10. R. Honsberger (ed.), *Mathematical Plums*, The Dolciani Mathematical Expositions, Vol. IV, M.A.A., Wash., D.C., 1979.
11. Z.A. Melzac, *Companion to Concrete Mathematics*, Vols. I, II, Wiley, N.Y., 1973, 1976.
12. G. Pólya, *How To Solve it*, Doubleday, N.Y., 1957.
13. G. Pólya, *Mathematical Discovery*, Vols. I, II, Wiley, N.Y., 1962, 1965.
14. G. Pólya, *Mathematics and Plausible Reasoning*, Vols. I, II, Princeton University Press, Princeton, 1954.

15. H. Rademacher and O. Toeplitz, *The Enjoyment of Mathematics*, Princeton University Press, Princeton, 1957.
16. A. W. Roberts and D. L. Varberg, *Faces of Mathematics*, Crowell, N.Y., 1978.
17. I. J. Schoenberg, *Mathematical Time Exposures*, M.A.A., Wash., D.C., 1983.
18. S. K. Stein, Mathematics, *The Man-Made Universe*, Freeman, San Francisco, 1976.
19. H. Steinhaus, *Mathematical Snapshots*, Oxford University Press, N.Y., 1969.

Problems:

20. G. L. Alexanderson, L. F. Klosinski, L. G. Larson, *The William Lowell Putnam Competition, Problems and Solutions*: 1965–1984, M.A.A., USA, 1985.
21. M. N. Aref and W. Wernick, *Problems and Solutions in Euclidean Geometry*, Dover, N.Y., 1968.
22. E. Barbeau, M. S. Klamkin, and W. O. Moser, *1001 Problems in High School Mathematics*, Vols. I–V, Canadian Mathematical Society, Ottawa, 1976–1985.
23. S. J. Bryant, G. E. Graham, and K. G. Wiley, *Nonroutine Problems in Algebra, Geometry, and Trigonometry*, McGraw-Hill, N.Y., 1965.
24. M. Charosh, *Mathematical Challenges*, N.C.T.M., Wash., D.C., 1965.
25. *Contest Problem Books* I, II, III, IV, Annual High School Mathematics Examinations 1950–60, 1961–65, 1966–72, 1973–82 (NML vols. 5, 17, 25, 29), M.A.A., Wash., D.C., 1961, 1966, 1973, 1983.
26. H. Dorrie, *100 Great Problems in Elementary Mathematics*, Dover, N.Y., 1965.
27. E. B. Dynkin et al, *Mathematical Problems: An Anthology*, Gordon and Breach, N.Y., 1969.
28. D. K. Fadeev and I. S. Sominski, *Problems in Higher Algebra*, Freeman, San Francisco, 1965.
29. A. M. Gleason, R. E. Greenwood, and L. M. Kelly, *The William Lowell Putnam Competition, Problems and Solutions*: 1938–1964, M.A.A., Wash., D.C., 1980.
30. K. Hardy and K. S. Williams, *The Green Book, 100 Practice Problems for undergraduate mathematics competitions*, Integer Press, Ottawa, 1985.
31. T. J. Hill, *Mathematical Challenges II Plus Six*, N.C.T.M., Wash., D.C., 1974.

32. A. P. Hillman and G. L. Alexanderson, *Algebra through Problem Solving*, Allyn and Bacon, Boston, 1966.

33. *Hungarian Problem Books* I and II (Eötvös Competitions 1894–1928), translated by E. Rapaport, NML vols. 11 and 12, M.A.A., Wash., D.C., 1963.

34. *International Mathematical Olympiads, 1959–1977*, compiled and with solutions by S. L. Greitzer, NML vol. 27, M.A.A., Wash., D.C., 1978.

35. G. Klambauer, *Problems and Propositions in Analysis*, Dekker, N.Y., 1979.

36. L. C. Larson, *Problem-Solving Through Problems*, Springer-Verlag, N.Y., 1983.

37. M. Lehtinen (ed.), *26th International Mathematical Olympiad, Results and Problems*, Helsinki, 1985.

38. L. Lovász, *Combinatorial Problems and Exercises*, North-Holland, Amsterdam, 1979.

39. F. Mosteller, *Fifty Challenging Problems in Probability with Solutions*, Addison-Wesley, Reading, 1965.

40. D. J. Newman, *A Problem Seminar*, Springer-Verlag, N.Y., 1982.

41. G. Pólya and J. Kilpatrick, *The Stanford Mathematics Problem Book*, Teachers College Press, N.Y., 1974.

42. G. Pólya and G. Szegö, *Problems and Theorems in Analysis*, Vols. I, II, Springer-Verlag, N.Y., 1976.

43. D. O. Shklarsky, N. N. Chentzov, and I. M. Yaglom, *Selected Problems and Theorems in Elementary Mathematics*, Mir, Moscow, 1979.

44. W. Sierpinski, *A Selection of Problems in the Theory of Numbers*, Pergamon, Oxford, 1964.

45. W. Sierpinski, *250 Problems in Elementary Number Theory*, American Elsevier, N.Y., 1970.

46. H. Steinhaus, *One Hundred Problems in Elementary Mathematics*, Basic Books, N.Y., 1964.

47. S. Straszewicz, *Mathematical Problems and Puzzles from the Polish Mathematical Olympiads*, Pergamon, Oxford, 1965.

48. G. Szasz et al (eds.), *Contests in Higher Mathematics, Hungary 1949–1961*, Akademiai Kiado, Budapest, 1968.

49. H. Tietze, *Famous Problems of Mathematics*, Greylock Press, N.Y., 1965.

50. I. Tomescu, *Problems in Combinatorics and Graph Theory*, Wiley, N.Y., 1985.

51. N. Y. Vilenkin, *Combinatorics*, Academic Press, N.Y., 1971.

52. I. M. Yaglom and V. G. Boltyanskii, *Convex Figures*, Holt, Rinehart and Winston, N.Y., 1961.

138   INTERNATIONAL MATHEMATICAL OLYMPIADS

53. A. M. Yaglom and I. M. Yaglom, *Challenging Mathematical Problems with Elementary Solutions*, Vols. I, II, Holden-Day, San Francisco, 1964, 1967.

Algebra:
54. S. Barnard and J. M. Child, *Higher Algebra*, Macmillan, London, 1939.
55. G. Chrystal, *Algebra*, Vol. I, II, Chelsea, N.Y., 1952.
56. C. V. Durell and A. Robson, *Advanced Algebra*, Vols. I, II, III, Bell, London, 1964.
57. H. S. Hall and S. R. Knight, *Higher Algebra*, Macmillan, London, 1932.
58. A. Mostowski and M. Stark, *Introduction to Higher Algebra*, Pergamon, Oxford, 1964.
59. J. V. Uspensky, *Theory of Equations*, McGraw-Hill, N.Y., 1945.

Combinatorics and Probability:
60. C. Berge, *Principles of Combinatorics*, Academic Press, N.Y., 1971.
61. R. A. Brualdi, *Introductory Combinatorics*, Elsevier North-Holland, N.Y., 1977.
62. D. I. A. Cohen, *Basic Techniques of Combinatorial Theory*, Wiley, N.Y., 1978.
63. L. Comptet, *Advanced Combinatorics*, Reidel, Dordrecht, 1974.
64. J. Riordan, *Introduction to Combinatorial Analysis*, Wiley, N.Y., 1958.
65. I. Tomescu, *Introduction to Combinatorics*, Collet's, Romania, 1975.
66. A. Tucker, *Applied Combinatorics*, Wiley, N.Y., 1980.
67. E. Borel, *Elements of the Theory of Probability*, Prentice-Hall, N.J., 1950.
68. W. Burnside, *Theory of Probability*, Dover, N.Y., 1959.
69. M. M. Eisen and C. A. Eisen, *Probability and its Applications*, Quantum, N.Y., 1975.
70. W. Feller, *An Introduction to Probability Theory and its Applications*, Vol. I, Wiley, N.Y., 1961.
71. M. A. Goldberg, *An Introduction to Probability Theory with Statistical Applications*, Plenum, N.Y., 1984.
72. S. Goldberg, *Probability, An Introduction*, Prentice-Hall, N.J., 1960.
73. F. Mosteller, R. E. K. Rourke, and G. B. Thomas, Jr., *Probability and Statistics*, Addison-Wesley, Reading, 1961.
74. J. V. Uspensky, *Introduction to Mathematical Probability*, McGraw-Hill, N.Y., 1937.
75. W. A. Whitworth, *Choice and Chance*, Hafner, N.Y., 1948.

Geometry (Plane):

76. P. Andreev and E. Shuvalova, *Geometry*, Mir, Moscow, 1974.
77. N. Altshiller-Court, *College Geometry*, Barnes and Noble, N.Y., 1952.
78. H. S. M. Coxeter, *Introduction to Geometry*, Wiley, N.Y., 1969.
79. D. R. David, *Modern College Geometry*, Addison-Wesley, Reading, 1949.
80. C. W. Dodge, *Euclidean Geometry and Transformations*, Addison-Wesley, Reading, 1972.
81. C. V. Durell, *Modern Geometry*, Macmillan, London, 1952.
82. C. V. Durell, *Projective Geometry*, Macmillan, London, 1952.
83. H. Eves, *A Survey of Geometry*, Vols. I, II, Allyn and Bacon, Boston, 1963.
84. H. G. Forder, *Higher Course Geometry*, Cambridge University Press, Cambridge, 1931.
85. L. I. Golovina and I. M. Yaglom, *Induction in Geometry*, Mir, Moscow, 1979.
86. R. A. Johnson, *Advanced Euclidean Geometry*, Dover, N.Y., 1960.
87. D. C. Kay, *College Geometry*, Holt, Rinehart & Winston, N.Y., 1969.
88. Z. A. Melzak, *Invitation to Geometry*, Wiley, N.Y., 1983.
89. D. Pedoe, *A Course of Geometry for Colleges & Universities*, Cambridge University Press, London, 1970.
90. G. Salmon, *A Treatise on Conic Sections*, Chelsea, N.Y., 1954.
91. P. F. Smith and A. S. Gale, *The Elements of Analytic Geometry*, Ginn, Boston, 1904.
92. N. Vasilev and V. Gutenmacher, *Straight Lines and Curves*, Mir, Moscow, 1980.
93. W. A. Wilson and J. I. Tracey, *Analytic Geometry*, Heath, Boston, 1937.

Geometry (Solid):

94. R. J. T. Bell, *An Elementary Treatise on Coordinate Geometry of Three Dimensions*, Macmillan, London, 1912.
95. P. M. Cohn, *Solid Geometry*, Routledge and Paul, London, 1965.
96. N. Altshiller-Court, *Modern Pure Solid Geometry*, Chelsea, N.Y., 1964.
97. A. Dresden, *Solid Analytical Geometry and Determinants*, Wiley, N.Y., 1930.
98. W. F. Kern and J. R. Bland, *Solid Mensuration with Proofs*, Wiley, N.Y., 1938.
99. L. Lines, *Solid Geometry*, Dover, N.Y., 1965.
100. G. Salmon, *A Treatise on the Analytic Geometry of Three Dimensions*, Chelsea, N.Y., 1954.

101. C. Smith, *An Elementary Treatise on Solid Geometry*, Macmillan, London, 1895.

Graph Theory:

102. B. Andrasfai, *Introductory Graph Theory*, Pergamon, N.Y., 1977.
103. B. Bollobas, *Graph Theory, An Introductory Course*, Springer-Verlag, N.Y., 1979.
104. J. A. Bondy and U. S. R. Murty, *Graph Theory with Applications*, American Elsevier, N.Y., 1976.
105. F. Harary, *Graph Theory*, Addison-Wesley, Reading, 1969.

Inequalities:

106. O. Bottema et al, *Geometric Inequalities*, Wolters-Noordhoff, Groningen, 1969.
107. G. H. Hardy, J. E. Littlewood, and G. Pólya, *Inequalities*, Cambridge University Press, Cambridge, 1934.
108. D. S. Mitrinovic, *Analytic Inequalities*, Springer-Verlag, Heidelberg, 1970.
109. D. S. Mitrinovic, *Elementary Inequalities*, Noordhoff, Groningen, 1964.

Theory of Numbers:

110. R. D. Carmichael, *The Theory of Numbers and Diophantine Equations*, Dover, N.Y., 1959.
111. G. H. Hardy and E. M. Wright, *Introduction to the Theory of Numbers*, Clarendon Press, Cambridge, 1954.
112. I. Niven and H. S. Zuckerman, *An Introduction to the Theory of Numbers*, Wiley, N.Y., 1960.
113. H. Rademacher, *Lectures on Elementary Number Theory*, Blaisdell, N.Y., 1964.
114. W. Sierpinski, *Elementary Theory of Numbers*, Hafner, N.Y., 1964.
115. J. V. Uspensky and M. A. Heaslet, *Elementary Number Theory*, McGraw-Hill, N.Y., 1939.

Trigonometry:

116. H. S. Carslaw, *Plane Trigonometry*, Macmillan, London, 1948.
117. C. V. Durell and A. Robson, *Advanced Trigonometry*, Bell, London, 1953.
118. E. W. Hobson, *A Treatise on Plane and Advanced Trigonometry*, Dover, N.Y., 1957.

Other:

119. E. R. Berlekamp, J. H. Conway, and R. K. Guy, *Winning Ways*, Vols. I, II, Academic Press, London, 1982.

120. M. P. Gaffney and L. A. Steen, *Annotated Bibliography of Expository Writing in the Mathematical Sciences*, M.A.A., Wash., D.C., 1976.

121. A. Gardner, *Infinite Processes-Background to Analysis*, Springer-Verlag, N.Y., 1982.

122. I. M. Gelfand, E. G. Glagoleva, and E. E. Shnol, *Functions and Graphs*, M.I.T. Press, Cambridge, 1969.

123. S. I. Gelfand et al, *Sequences, Combinations, Limits*, M.I.T. Press, Cambridge, 1969.

124. W. Gellert et al (eds.), *The VNR Concise Encyclopedia of Mathematics*, Van Nostrand Reinhold, N.Y., 1977.

125. S. Goldberg, *Introduction to Difference Equations*, Wiley, N.Y., 1958.

126. I. Niven, *Maxima and Minima Without Calculus*, The Dolciani Mathematical Expositions, Vol. VI, M.A.A., Wash., D.C., 1981.

127. W. L. Schaaf, *Bibliography of Recreational Mathematics*, Vols. I, II, III, IV, N.C.T.M., Wash., D.C., 1965, 1972, 1973, 1978.

128. S. Schuster, *Elementary Vector Geometry*, Wiley, N.Y., 1962.

Also:

The entire collection of the New Mathematical Library (now 31 volumes —see page vi of this book), available from The Mathematical Association of America, is highly recommended.

Here is an alternative solution to problem **N/2**, by Arthur Engel, added in the second printing of this book:

If any x_i changes sign, the sum does not change modulo 4. Now change, one by one, every -1 to $+1$ so that the sum becomes n. Since the sum must remain $\equiv 0 \pmod 4$, so must $n \equiv 0 \pmod 4$.